Student Study Guide for use with

Biology for the Informed Citizen with Physiology

Sharon L. Gilman

New York   Oxford

Oxford University Press

2014

Oxford University Press is a department of the University of Oxford.
It furthers the University's objective of excellence in research, scholarship,
and education by publishing worldwide.

Oxford   New York
Auckland   Cape Town   Dar es Salaam   Hong Kong   Karachi
Kuala Lumpur   Madrid   Melbourne   Mexico City   Nairobi
New Delhi   Shanghai   Taipei   Toronto

With offices in
Argentina Austria Brazil Chile Czech Republic France Greece
Guatemala Hungary Italy Japan Poland Portugal Singapore
South Korea Switzerland Thailand Turkey Ukraine Vietnam

For titles covered by Section 112 of the US Higher Education Opportunity
Act, please visit www.oup.com/us/he for the latest information about
pricing and alternate formats.

Published by Oxford University Press.
198 Madison Avenue, New York, NY 10016
www.oup.com

Oxford is a registered trademark of Oxford University Press

ISBN 978-0-19-995801-6

Printing number: 9 8 7 6 5 4 3

Printed in the United States of America
on acid-free paper

Contents

Student Study Guide

Chapter 1 The Nature of Biology and Evolution

Why Does Biology Matter to You?

You are a living thing on an earth full of other, interesting living things, and biology is all about all of us. Why are we the way we are? How do we do what we do? What are the opportunities, risks, and moral dilemmas raised by the rapid advances in biological research going on now? To understand all this and deal with the challenges, you need to understand the science behind them. That is the goal of this book.

1. Case Study: The Infidelity Gene

1. Case Study Summary

John and Mary Smith, a young married couple, decide to start a family and elect to have their own DNA screened for possible genetic issues that could affect their child's health and development. All is well, except John is found to have the "infidelity gene," *AVPR1a.* (Names of human genes are always capitalized and placed in italics.) As it turns out, *AVPR1a* controls production of a protein called the *arginine vasopressin receptor* which is found in membranes of nerve cells in the brain and is responsible for transporting the hormone *arginine* vasopressin from the bloodstream to the nerve cells. There are several versions, or **alleles**, of this gene in humans, and *AVPR1a* makes less protein than the others, so John's brain gets less of the hormone. Less of this hormone may result in less interest in one's mate, which could lead to a higher likelihood of John cheating on Mary.

What are hormones, alleles, receptors? What does a protein have to do with adultery? Mary (and you) clearly need to learn some biology.

Questions for Review:

How does the "infidelity gene" possibly cause a person to cheat on his or her spouse?

Learning Objective 1.1: How Does Biology Affect Your Life?

Explain how understanding biology can help you make informed decisions as a citizen.

1.1   Summary

As you see with Mary, learning a little biology raises more questions than answers. Should you get genetic testing? What does it tell you? What should you do with that information? Why do you have the genes you do, and why do some genes seem to do more harm than good? The better you understand biology, the more informed decisions you can make.

The science of biology relates to questions of personal health, the role of diet and exercise, injuries, diseases and treatments, and eventually questions about aging, cancer, cardiovascular problems, and end of life choices. Researchers continue to discover new information about all of these. Luckily, there will be doctors and counselors to help you keep up, but understanding the biology behind the choices will help you make the best decisions.

Biology isn't just about you though. Issues from science and biology raise questions that affect your public life and the planet. You can get good information from the internet, newspapers, television, blogs and books, but in the end you have to evaluate what you learn and make responsible decisions for your life and the lives of others. And biological research keeps advancing. An understanding of biology will help you make better informed decisions in many areas of your life.

In addition to all that, biology is just interesting. Life comes in every size, shape, color, lives nearly everywhere, and has endless ways of living. Diversity of life is fascinating, but living organisms also have many features in common. Understanding this diversity and commonality is what evolution is all about.

1.1 Questions for Review

1.1.1 List a few examples of personal decisions you might have to make which have to do with biology.

1.1.2 List a few examples of societal issues you might have to consider which have to do with biology.

Learning Objective 1.2: What are the Features of Life

Use the four principle features of life to discriminate living from non-living things.

1.2 Summary

Living things have four features in common:

1. Metabolism: All living organisms take in energy and food and use them to respond to the environment, survive, and reproduce.

2. Inheritance and reproduction: All living organisms can reproduce, and each inherits biological information from its parents or parent.

3. Evolution: All living organisms are related and all species change over time. Evolution is the central feature of life because all the other features are the result of evolutionary change.

4. Diversity: All living things have evolved into distinct, unique forms, and this diversity contains patterns that confirm life's evolutionary past.

1.2 Questions for Review

1.2.1 What are the four features of living organisms?

1.2.2 Explain why is evolution the central feature of the four.

Learning Objective 1.3: How Do Organisms Function?

Describe the relationship among atoms, macromolecules, and cells, and explain how enzymes maintain cell function.

1.3 Summary

One of life's wonders is that simple chemical atoms can be arranged into complicated structures capable of contributing to the complex characteristics and behaviors of living things. To understand how organisms are put together, function, and behave, you have to understand some chemistry.

Atoms, Chemical Bonding, and Molecules

Atoms are the building blocks of every physical thing. They are composed of electrons orbiting a central sphere or nucleus made of protons and neutrons. Protons are positively charged, neutrons are neutral, and electrons are negatively charged. Elements are unique combinations of atoms that can't be further broken down.

Four elements account for close to 96% of all atoms in living organisms: carbon, oxygen, hydrogen, and nitrogen.

An atom is either reactive or inert. The outermost orbital or shell of an atom can accommodate a specific number of electrons. When the shell is filled, the atom is unreactive or inert. If an atom needs just one electron to complete its outer shell, or if it has only one electron in its outer shell, it will be highly reactive, combining with other atoms to either gain or lose an electron. Sodium and chlorine, for example, are like this. They need to lose and gain one electron, respectively, so they easily react with one another to form sodium chloride, or table salt.

Many atoms have partially filled outer shells such that they need to share orbits with other atoms to have complete shells. This sharing of electrons is chemical bonding which combines different elements into molecules. For example, carbon and oxygen chemically bond to each other to form carbon dioxide.

Macromolecules

Macromolecules are large molecules made up of many atoms. Carbon-based macromolecules and water make up most of the fundamental unit of life, the cell. There are four of these macromolecules:

1. Nucleic Acids: DNA is one type of nucleic acid. Nucleic acids are made up of nucleotides.

2. Proteins: Proteins are composed of amino acids. They regulate chemical reactions in cells, and they make up many of the parts of cells and organisms.

3. Carbohydrates: These are sugars, like glucose, joined together. Cellulose is the structural carbohydrate in plant cell walls. Starch is a carbohydrate that is a good food source.

4. Lipids: Fatty acids and glycerol join together to make lipids. These are an excellent energy source and also serve as insulation and shock absorbers for organs.

Cells

Living things are made up of and maintained by small compartments called cells. Cells are enclosed by a cell membrane, which controls what can enter and leave the cell. The cells contain water-soluble molecules in solution, and fats, which do not dissolve in water. Neither can pass across a cell membrane unassisted.

Chemical Reactions and Enzymes

In cells and organisms, molecules are taken apart to produce smaller components, or they are joined together to make larger structures. Some of these events require large inputs of energy, and sometimes they release energy, but usually only with some assistance. That's where enzymes come in. So, for example, putting nucleotides together to form DNA is a biosynthetic reaction that requires energy input. Breakdown of starch, however, releases energy, but only after an enzyme has done its work.

Enzymes are proteins that make it easier for chemical reactions to occur. They lower the amount of "start-up" energy required in chemical reactions in cells and organisms.

Energy Extraction and Use

Metabolism is all the chemical reactions in cells that capture, extract, convert, and use energy. Ultimately that energy has come from the sun, which plants are able to convert to chemical bond energy in sugars. Organisms then extract that energy from food to use it for functions required for living. The metabolic reactions are regulated by specific enzymes. Without the enzymes, the reactions won't happen.

1.3 Define these Key Terms

atom

electron

element

chemical boding

molecules

macromolecules

cell

enzymes

1.3 Questions for Review

1.3.1 Label the parts of an atom in Figure 1.5.

1.3.2 Living things are mostly made up of what four elements?

1.3.3 Fill in the following table.

Macromolecule	Subunits	Functions

1.3.4 What is a cell?

1.3.5 Where does the energy for metabolism come from?

1.3.6 How do enzymes make metabolism possible?

Learning Objective 1.4: How Do Organisms Reproduce?

Describe the relationship between DNA, genes, and alleles, and compare how genes are inherited in sexual and asexual reproduction.

1.4 Summary

In the case study, John almost certainly inherited the infidelity allele. How does that work?

Inheritance

Organisms like us inherit from both our mother and our father. Each parent has a complete set of genes but different combinations of alleles for those genes. This means you inherit a complete set of human genes, but your combination of alleles is unique (unless you have an identical twin). Inheritance is this passing of information from one generation to the next. These genes that get passed along from one generation to the next contain all the necessary instructions to grow, live, and reproduce.

Reproduction

We have to reproduce because we can't live forever; our parts wear out and accidents happen, and habitat changes. Death is a universal feature of life. The steps in reproduction ensure that each new individual inherits a full complement of genetic information stored in

DNA. The chemical properties of DNA allow it to both store information and pass it along to the next generation. This occurs in two ways.

1. Asexual Reproduction

This is the simplest form of reproduction, occurring in individual cells like those in your body or in a single-celled organism like bacteria. DNA is copied, so there are two complete sets. The cell divides and each daughter cell gets a complete DNA copy. Therefore the daughter cells are just about identical to the parent. There's no mother, father, or sex involved.

The daughter cells are not identical to the parent though, because while the DNA copying mechanism is quite accurate, it's not perfect. Mistakes called mutations get made. When a mutation occurs in a gene, it makes a new allele of that gene. There's a very slight chance that John got his copy of *AVPR1a* from a mutation rather than a parent, but it's unlikely because mutations are so rare.

2. Sexual Reproduction

Sexual reproduction occurs when an egg from the mother fuses with the sperm cell from the father to form a new individual. In this case, the parent's DNA was not just copied (like in asexual reproduction) but was combined to make a new, unique individual.

Without reproduction, life would cease to exist, so it's a critical part of life. Two facts about the processes of reproduction and inheritance are especially noteworthy:

> 1. Living organisms are part of an unbroken chain of ancestors and descendants, extending back into time. Reproduction divides this chain into generations, and inheritance guarantees some degree of genetic continuity. Ever since life got started, new life has come only from existing life, not from nonliving sources.

> 2. The combination of mutation and sexual shuffling of alleles guarantees that individuals in one generation differ from each other and from the previous generation. This genetic variation is the basis for life's most central feature: evolution.

1.4 Define these Key Terms

alleles

asexual reproduction

sexual reproduction

mutations

inheritance

1.4 Questions for Review

1.4.1 Draw flow charts of the steps in both sexual and asexual reproduction. What is different about the resulting "offspring?"

1.4.2 Explain how new alleles might arise from genes

Learning Objective 1.5: How Does Life Evolve?

Create a scenario describing how a population of organisms might adapt to a new environment.

1.5 Summary

John likely inherited the infidelity gene from a parent, but how did they get it? Humans have about 25,000 genes and most have several alleles. Some alleles are common in certain populations. For example, blonde hair and blue eyes in Scandinavia, and black hair and dark eyes in Africa. The chances of inheriting a particular allele depend on how common it is in your particular population.

Frequencies of alleles changes over time. If a lot of people have it, the frequency goes up. If not many have it, the frequency goes down. Changes in frequency of alleles reflect a *genetic change in the population*, the modern definition of evolution.

Darwin's Theory

In 1859, when Darwin proposed his theory of evolution, he didn't know about DNA, genes or alleles, but he did know about inheritance, and he knew that

- o Individuals in a population vary from each other.
- o At least some of these variations are inherited.
- o In every population, many or most offspring die before reaching maturity.

Darwin asked a simple question: which offspring die? He reasoned that certain offspring survive better because they have some advantage—they're faster, bigger, better able to hide—something gives them an edge. The offspring that lack these variations die. The ones that survive are the ones that reproduce the next generation and so their traits are inherited and the beneficial variations become more common.

Darwin called this "descent with modification." Species change over time. Darwin used the term **natural selection** to describe how these changes occur—the way that environmental conditions "selected" which individuals would survive and reproduce.

Extending Darwin's Theory Through Time

Like any "big" idea in science, Darwin's idea that species can change had important ramifications:

1. As species change, they become better adapted to their environment, which is important because in Darwin's time geologists were showing that conditions on earth changed over time. If a species couldn't change as the environment changed, it would go extinct.

2. If a species changed enough, it might change into a different species. If a single species had populations living in different areas and those areas changed independently of one another, the populations in the different areas would change independently too. Over time they would become increasingly different from one another until there were two distinct species where there had been just one.

3. If we run Darwin's process backward through time, instead of species splitting apart, they merge into common ancestors. Because all life forms are part of a continuous chain of reproduction, the logical conclusion is that all species are related—an ever expanding tree of life.

Evolution as the Unifying Theme of Biology

Evolution says that all forms of life are related. This can be difficult to see at first, but that idea explains so many facts of nature that biologists see evolution as the central theme that ties together all of biology.

For example, consider us. We are vertebrates: animals with internal skeletons made of bone with well-defined head that contain our brain and sensory organs. Because vertebrates range in size from guppies to whales, live just about everywhere from the equator to the poles, and have widely varying lifestyles, you might expect their bodies would be quite different too. But look at our body plan. A skull attached to a vertebral column that supports ribs, two pairs of limbs, and ends in a tail. Those bones vary in size and shape, and are lost in some vertebrates, but mostly the bones are the same, with the same muscles that move them. Our organs are in about the same place and we have similarly organized blood vessels and nerves.

It is evident that all vertebrates are variations on a theme rather than new designs. This is exactly what evolution predicts. Vertebrates inherited their similarities from common ancestors and evolved differences as they adapted to environments. It's difficult to explain this pattern of diversity and similarity any other way.

1.5 Define these Key Terms

natural selection

evolution

1.5 Questions for Review

1.5.1 Explain why a particular allele would have a high frequency in a population.

1.5.2 If an allele is in high frequency in a population, what would tend to happen to the frequency in subsequent generations if conditions remained the same?

1.5.3 What would happen in question 1.5.2 if conditions changed?

1.5.4 What is the definition of evolution?

1.5.5. Explain how natural selection can result in two new species from one.

1.5.6 Explain how the group "vertebrates" illustrates the idea that we're all related through common ancestors.

Learning Objective 1.6: What Patterns of Diversity are Found in Nature?

Explain the connection between common ancestors, the diversity of environments on earth, and the scientific classification system.

1.6 Summary

Scientists have described about 1.3 million species. There may be 30 million living species and 50 billion have already gone extinct. Life is clearly quite diverse. This diversity raises two issues for science:

> Why does this diversity exist and how can it be organized in a meaningful way, reflecting the underlying cause?

Life is Diverse

Our planet is full of different habitats offering different challenges to life, and it also changes all the time. Living things constantly have to change in order to get what they need and keep up. This is how new species arise. New species arise when older species adapt to the new and different living conditions.

Evolutionary Diversification Leads to Degrees of Relatedness

Diversity on Earth has a distinct pattern. Consider your own family tree, back a few generations. The tree starts with some distant great-great-grandparent. All his or her children come next, and all their children next, until we get to you. Your tree includes your parents, siblings, aunts, uncles, and cousins. You're related to all of them. You're related most closely to your parents and siblings, so you probably look most like them.

This is the same pattern in the tree of life on Earth. Closely related species share a recent common ancestor (like your siblings share your parents), and they have much in common. If two species are just in the process of splitting apart, it might be hard to tell them apart. The farther back you go in time, the fewer characteristics the species share. The tree is arranged in a **hierarchy**, just like your family tree. Evolution predicts that we should see this hierarchy of relatedness, and we do.

Organizing Hierarchies in the Diversity of Life

Species lie at one end of life's hierarchy and are so genetically similar that they can interbreed successfully. Dogs, for example, come in many shapes and sizes, but any dog can breed with any other dog because they're all dogs. Dogs are part of a bigger group that includes wolves and coyotes. They're carnivores, which puts them in a larger group that includes cats and seals. These are all mammals, and all mammals are vertebrates. Natural diversity consists of a hierarchical organization of groups within groups.

This hierarchy was formalized by Carolus Linneaus in the early 1700s. He came up with the following scheme:

Kingdom

    Phylum

        Class

            Order

                Family

                    Genus

                        Species

The species is the smallest group, and as you go up levels, the groups include progressively more organisms. Every organism can be classified according to this system. And its classification is, in fact, its evolutionary family tree. The fact that all organisms can be classified and organized in this way is strong evidence for evolution and the origin of species.

As a result of this hierarchical classification system, each organism can be assigned a unique scientific name, including its genus and species names. The words are Latin so they are italicized, and the genus is capitalized.

1.6 Define these Key Terms

diversity

hierarchy

scientific name

1.6 Questions for Review

1.6.1 Explain how the "tree of life" on Earth is like any other family tree.

1.6.2 Explain how the fact that life on Earth can be organized into a hierarchical classification system is evidence of evolution (or evidence that we're all related).

1.6.3 You're a biologist who has just discovered a new species in a new genus of beetles. Write an appropriate scientific name for this creature.

In every chapter in your text, the following five sections appear as special boxes within the chapter. You should refer to those as you try to answer the review questions here, which focus on the main point(s) of the section.

1. Biology in Perspective

Why does biology matter? Biology can be important in personal decisions, health decisions, decisions about controversial issues, and ethical questions. It also makes you aware of how much you have in common with every other living thing, and it gives you a clearer understanding of how the fascinating world works.

Question to consider:

What are a few of the reasons you should learn some biology?

1. Scientist Spotlight: Carol W. Greider (1961-)

Carol Greider began doing research as an undergraduate at the University of California at Santa Barbara. She got excellent grades, research experience, and letters of recommendation, but due to dyslexia, she did poorly on her GREs and almost didn't get into graduate school. But finally, she did get into Elizabeth Blackburn's lab at U.C. Berkley.

Together, Carol Greider and Elizabeth Blackburn discovered a molecule called telomerase. It helps prevent some of the effects of aging in a cell, allowing the cell to divide over and over again. Since then, Greider has shown that as cells age, telomeres shorten, and once they get too short, they lose their protective power, chromosomes become damaged, and cells die, and that cancer cells have overactive telomerase and telomeres that are too long.

Question to consider:

What does telomerase do that makes its discovery worthy of a Nobel Prize?

1. Technology Connection: Identifying the "Infidelity Gene"

The infidelity gene is real, although it's not clear that if affects humans. The *AVPR1a* gene controls the transport of the hormone arginine vasopressin from the blood to the brain. Sexual activity and other social interactions between couples release the hormone which makes you feel good, possibly strengthening the pair bond. This has been studied in voles (mouse-like animals), however, not humans. Prairie voles are monogamous (one male and one female form a pair bond and mate only with each other), while meadow and montane voles are promiscuous (both sexes mate with different partners and do not form pair bonds). It turns out that there are different alleles of the *AVPR1a* gene in the monogamous and promiscuous species. The popular press picked up the story and labeled the allele in the promiscuous species the "infidelity gene." Science depends on big sample sizes and repeated tests, and these studies so far have neither. An *association* between an allele and a behavior doesn't prove the allele *causes* the behavior.

Question to consider:

What do scientists <u>actually</u> know about the *AVPR1a* allele?

1. Life Application: Determining When Life Has Ended

It used to be that when a person's heart stopped beating they were considered dead. By 1980 the Uniform Determination of Death Act (UDDA) was drafted to define clear criteria for neurological or brain death. The UDDA states "An individual who has sustained either (1) irreversible cessation of circulatory and respiratory functions, or (2) irreversible cessation of all functions of the entire brain, including the brainstem, is dead." This was to serve as a model for state laws.

Based on state guidelines, hospitals have strict protocols for determining brain death.

Question to consider:

How do physicians determine that a person is brain dead?

1. How Do We Know?: Spontaneous Generation

Before we knew that the only way individuals arise is from their parents, through reproduction, people thought that life could appear spontaneously.   This issue was put to rest by Francesco Redi in 1668.

But many people still thought that microorganisms like bacteria could do so, because they were so small and simple. In 1859, the French Academy of Sciences (FAS) offered a prize for the best experiment proving or disproving spontaneous generation. French microbiologist and chemist, Louis Pasteur entered the contest and won.

Question to consider:

How did Redi, and later, Pasteur, demonstrate that organisms can only come from other organisms, not spontaneous generation?

Student Study Guide

Chapter 2 The Nature of Science

How Do We Know How the World Works?

The world appears to work in a regular, repeatable way that enables scientists to find the "rules" governing physical and biological processes on Earth. Scientists may try to solve a particular problem, answer a specific question, or determine how certain processes work. Sometimes this is just out of curiosity, and sometimes there are important consequences impacting medicine, industry or society. A few famous scientists (Newton, Einstein, Darwin, for example) have developed comprehensive theories that explain a great deal.

2. Case Study: The Mysterious Case of Childbed Fever

2. Case Study Summary

In 1846, a woman in labor, about to give birth, arrived at the hospital. She had heard that it was better to go to the section with the midwives, rather than with the doctors, but she figured that must be a rumor and surely doctors were better qualified to help her successfully give birth. She gave birth to a baby boy, but a few hours later she began to feel discomfort in her abdomen. Over the next hours, the pain worsened, she became nauseous, vomited, developed a high fever, and her abdomen filled with gas. In three days she was dead.

An autopsy revealed a swollen uterine wall, foul smelling gases, pus and putrid flesh. This was childbed fever, a disease that claimed the life of one out of every ten women who delivered babies with doctors in the 1840s. What was wrong and what has changed?

Learning Objective 1: How Would a Scientist Investigate Childbed Fever?

Outline the steps that Ignaz Semmelweis took to find the cause of childbed fever.

2.1   Summary

Dr. Ignaz Semmelweis was the director of a maternity division in 1846, a year when 459 women coming through his division died of childbed fever. He became obsessed with figuring out the cause and prevention of the disease. Because he was a scientist, the first thing he did was collect data. He compiled birth and death records at the hospital due to childbed fever. Death rates were 2.5 times higher in Division 1, where the male doctors and their male students worked, compared to Division 2 where female midwives and their female students treated the women. He also knew that women who delivered babies outside the hospital hardly ever contracted childbed fever. The problem must lie with hospital practices.

1. Miasma? Semmelweis first looked at miasma as a cause. Miasma was a harmful, toxic vapor supposedly exhaled by sick people or exuded by garbage or sewers. It was widely believed that if you breathed this in, you could become sick. If this were the problem, perhaps improved ventilation at the hospital would help? This did not help, and also didn't make much sense because Division 1 and Division 2 shared the same air.

2. Delivery Method? The doctors had the mothers lie on their backs during labor, and the midwives had them lie on their sides. He instructed the doctors to change their practice, but it made no difference.

3. Were the male medical students being too rough on the patients? Semmelweis reduced the number of medical students by half and reduced the number of exams the patient underwent. This made no difference.

4. What else? A frustrated Semmelweis went on vacation in 1847 and when he returned, he learned of the death of a friend and fellow doctor, Jakob Kolletschka. As it turned out, the doctor had been stabbed in the finger by a medical student's scalpel during an autopsy of a woman who had died of childbed fever, and he died childbed fever. Semmelweis concluded that he must have contracted "cadaverous particles" from corpse of the victim of childbed fever. We'd call those germs or microorganisms (like bacteria).

He was correct. Only the doctors and medical students did autopsies on the women who died of childbed fever, and they tended to do these in the morning prior to working with patients in the afternoon, and nobody sterilized anything at that time. They were passing the childbed fever germs right to their patients, in a way the midwives were not.

Semmelweis required all medical personnel to thoroughly wash their hands and scrub under their nails with chlorine bleach before entering the ward to examine women or deliver babies. Deaths from childbed fever in Division 1 dropped to just 3% of women, the same as in Division 2.

At this time, doctors knew microorganisms existed, but most did not believe they caused disease, so these practices were not widely adopted until 25 years later when "Germ Theory" linked microorganisms and disease. (See "Scientist Spotlight).

2.1 Define these Key Terms

miasma

microorganisms

2.1 Questions for Review

2.1.1 Draw a flow chart beginning with Dr. Semmelweis's observation of a problem and ending with his fixing it. What were the steps involved?

Learning Objective 2.2: How Does Science Work?

Determine whether a given statement is an observation, hypothesis, or theory, and use the steps of the scientific method to design and evaluate a simple experiment.

2.2 Summary

Scientists like Semmelweis follow particular steps to investigate a problem. This is called the scientific method. It includes several parts:

1. Observations and Facts

Observations are what you can hear, smell, taste, or feel physically. Facts are thing you know to be true about the natural world. Scientists want to explain the observations and facts they see. Semmelweis observed two facts: 1) deaths due to childbed fever happened more often in the hospital than outside and 2) those deaths happened more in Division 1 than Division 2.

2. Hypotheses and Predictions

A hypothesis is a possible cause or mechanism that could explain the observations and facts. Semmelweis came up with several: miasma, birthing position, rough treatment.

A hypothesis leads to a prediction about an outcome. Semmelweis had a prediction for each of his hypothesis. *If* miasma caused childbed fever, *then* ventilation will reduce the incidence of the disease. *If* the hypothesis is correct, *then* under certain conditions you should get certain results.

3. Testing

To test a hypothesis, a procedure is designed to set up the conditions the predictions require. If the predicted results occur, the hypothesis may be correct. If they don't the hypothesis is probably wrong. Semmelweis tested all his hypotheses. He added ventilation, he changed the birthing position, he reduced examinations by male students, and in each case he predicted that the incidence of childbed fever would go down and it didn't. His hypotheses were not supported by his results. But when he predicted that handwashing to remove particles would reduce the number of cases, it did. This hypothesis was supported.

Many tests take the form of controlled experiments. All factors which could affect the outcome of the test are controlled, or held constant, except for the one under investigation. For example, if you wanted to test whether fertilizer helped plants grow, you'd add fertilizer to some plants but not to others, AND you'd keep everything else about the plants the same: plant type, water, soil, light, temperature. Then if you saw a difference in plant growth, it must have been from the fertilizer.

4. Evaluation and Interpretation of Results

The scientist looks at whether the outcome is what was predicted by the hypothesis, but then needs to repeat the experiment to be sure. If the predicted result did not occur, maybe there was a mistake in performing the experiment. If the predicted results still don't happen, probably the hypothesis is wrong and the scientist must come up with a new one and repeat the whole process.

If the results do match the predictions, the hypothesis is supported, but the scientist still needs to retest to make sure those results are consistent. And most hypotheses lead to more than one set of predictions, so those need to be tested to make the case for the hypothesis stronger. The more evidence there is to support the hypothesis, the easier it is to convince other scientists the hypothesis is correct. But science is never finished. A hypothesis is never proven with absolute certainty. The next experiment might refute it, or, because nature is complex, you might not have the correct cause for your results. This does not mean that science is never correct. What it means is that scientific ideas are always open to be replaced by even more accurate ideas as more information is learned. Computers and medical care, for example, keep getting better as we learn more science.

Scientific Theories

There is Isaac Newton's theory of gravity, Charles Darwin's theory of evolution by natural selection, and Albert Einstein's theory of relativity. What makes these different from other scientific ideas such that the people behind them are famous?

Theories are similar to hypotheses, except that they look at the bigger picture of how nature works and encompass several fields of science. Both make predictions that can be tested, but hypotheses focus on explanations of smaller, specific events, while theories address a broad spectrum of events. Hypotheses are puzzle pieces and theories are the way the pieces fit together.

For example, the predictions of the theory of evolution by natural selection have been tested by geologists, paleontologists, and cosmologists, in addition to biologists. Evidence from studies in all these disciplines so far support the theory. So a theory is as close to proof as you get in science, although there is always room for new information, just like with an individual hypothesis. A theory demonstrates that nature operates according to rules we can discover and understand.

2.2 Define these Key Terms

scientific method

observations

facts

hypotheses

predictions

controlled experiments

scientific theory

2.2 Questions for Review

2.2.1 Indicate whether each of the following is an observation, hypothesis, or theory:

a. It gets colder on winter nights when the sky is clear rather than cloudy.

b. Everything with mass has a gravitational force.

c. Your heart rate increases when you exercise.

d. The more you exercise, the higher your heart rate will get (up to a point).

e. Whether hydrangea flowers are blue or pink depends on the soil pH.

f. Animals are well adapted to their environment because of natural selection.

2.2.2 You observe that your car won't start. Design a simple experiment to solve the problem. Include all the parts of the scientific method (observation, fact, hypothesis, prediction, test) and what result would support your hypothesis.

Learning Objective 2.3: What Assumptions Does Science Make About Nature?

Use the assumptions of science to determine whether a statement is scientific.

2.3 Summary

Based on hundreds of years of study of the natural world, there are four principles of science: cause and effect, consistency, repeatability, and materialism.

1) Cause and Effect. Every event or outcome in nature has a cause or source. A scientist can learn about causes by observing effects. And the reverse is also true: if a scientists sets up the correct causes, the effects are predictable.

2) Consistency and 3) Repeatability. Science assumes events are consistent and repeatable. If the same causes or conditions are set up, the same effects or results will occur. This is why scientists repeat their own experiments and the experiments of others. If results cannot be repeated, they are suspect. In this way science is self-correcting. Mistaken results get tossed.

4) Materialism. Effects in the natural world have natural, not supernatural, causes. Events with supernatural causes need not follow the rules of science, which makes them untestable

by science. This can put other approaches to knowledge, like religion, in conflict with science, but it's just something science cannot study. One either believes it or not.

2.3 Define these Key Terms

cause and effect

consistency

repeatability

materialism

2.3 Questions for Review

2.3.1 Explain what the following assumptions of science are, using Semmelweis's studies of childbed fever as an example:

> a. Cause and effect

> b. Consistency and repeatability

> c. Materialism

2.3.1 Explain why a, b, and c are each necessary for science to work (what couldn't scientists do or what would they do differently if these conditions were not assumed to hold?)

Learning Objective 2.4: What are the Principal Features of Science?

Explain the principal features of science.

2.4 Summary

Along with assumptions described above, there are three principal features of science.

1. Empirical Evidence. Science is based on evidence, not opinion or belief. Empirical evidence is information you get from direct observation, from experience, or from experimental results. Every scientific hypothesis and theory is evaluated strictly by how well it explains the existing empirical evidence. It doesn't matter if it seems logical or is popular. If it doesn't fit the evidence, it's wrong.

2. Testability. For a question or hypothesis to be addressed, it must be testable. There must be a way to determine evidence in support, either by experiment or observation. It must be falsifiable, or able to be proven wrong. If you don't get the results you predicted based on your hypothesis, your idea is wrong (Or else your experiment wasn't set up correctly. Either way, it's back to the drawing board.) False hypotheses are eventually identified and discarded, leaving only those hypotheses supported by empirical results.

3. Generality. Scientific investigations often apply to just a particular situation or process, but scientists are also interested in the generality of their work, or how widely it applies in other situations. When they publish papers, they point out how their work relates to that of other scientists. Gradually, they develop an understanding of the general principles that underlie their individual results, and these principles combine into broad scientific theories about how the world works.

2.4 Define these Key Terms

empirical evidence

testability

falsifiable

generality

2.4 Questions for Review

2.4.1 Describe how each of these three principles of science relate to Dr. Semmelweis's work with childbed fever.

a. Empirical evidence

b. Testability

c. Generality

Learning Objective 2.5: How Does Science Differ from Other Ways of Knowing?

Distinguish science from other ways of knowing.

2.5 Summary

Science explores particular questions about the natural world with a specific set of practices. It is not necessarily superior to other ways of thinking. There are some questions that science is not equipped to help answer, and other areas of knowledge address those. Theology seeks to understand God. Philosophy explores what it means to have a good life. Scientists aren't going to answer questions about spiritual domains and religious scripture isn't the place to go to learn about the natural world. Some view this as a conflict, but science and religion generally seek answers to different questions, so it need not be.

There is a great deal of overlap between science and some academic fields. Psychologists have a great deal in common with biologists who study brain features. Historians may work with evolutionary biologists in looking at fossils.

2.5 Questions for Review

2.1 For each of the following statements, state which of the following fields offers the best way to approach the topic: science, religion, or philosophy?

a. Define "the pursuit of happiness."

b. Moss grows on the north side of trees.

c. When you die you go to heaven.

d. What is truth?

e. Do prayers get answered?

f. Hand washing helps keep people from catching a cold virus.

2.2 For each statement above that can't be addressed by science, explain why not.

Learning Objective 5.6: How Does Science Differ from Pseudoscience and Quackery?

Determine whether a claim is likely science, pseudoscience, or quackery.

2.6 Summary

Just as important as understanding what science is, is the ability to recognize bad science, or pseudoscience or quackery. Quackery is promoting the use of a product or remedy when there is no plausible rationale for its effectiveness. Pseudoscience attempts to look like actual science so its assertions might appear valid. It begins with a claim and then presents only things that support that claim. Controlled experiments or direct tests are not done.

Instead of considering facts, pseudoscience appeals to emotion, sentiment, or distrust of established knowledge. In pseudoscience, nothing is ever revised or learned.

One pervasive example of pseudoscience is creationism or the idea that a divine being created all species on Earth in a single event. A person may accept this as faith, if they take the Bible literally, but it isn't science. Creationism can't be tested. Evidence against it is ignored: species have clearly changed through time and the Earth is very old. To deny this is to ignore the geologic record, the fossil record, evidence from embryos, and molecular studies.

Quackery is even worse because it takes advantage of people who are sick and vulnerable by offering "cures" or supplements that cause weight loss, for example, or bracelets that reduce arthritis pain. These ads are everywhere. Your best defense is to be skeptical and to know some biology.

2.6 Define these Key Terms

pseudoscience

quackery

2.6 Questions for Review

2.6.1 Explain and give an example of pseudoscience.

2.6.2 Explain and give an example of quackery.

2.6.3 Explain how pseudoscience is not science.

In every chapter in your text, the following five sections appear as special boxes within the chapter. You should refer to those as you try to answer the review questions here, which focus on the main point(s) of the section.

2. Biology in Perspective

The scientific method of inductive reasoning to develop and test hypotheses is the best way to learn about the world of nature. Scientific theories explain broad areas of science and fundamental natural laws. It is critical for citizens to understand how science works in order to appreciate its value to society, but also to recognize and reject pseudoscience and quackery.

Questions to Review

Explain the process of the scientific method.

Why is it important for a non-scientist citizen to know something about science?

2. Scientist Spotlight: Robert Koch (1843–1910)

Robert Koch wanted to show scientifically that the anthrax bacterium *caused* the disease *anthrax*. He tested this by injecting mice with blood from the spleens of animals that had died of anthrax. He compared these to mice injected with blood from healthy animals and found that only the mice injected with the anthrax bacteria died. Koch then developed pure cultures of the anthrax and found that these produced spores when growth conditions were poor. The spores could then germinate to create new bacterial cells when conditions were good. Koch traveled the world studying the infectious origins of diseases including tuberculosis, cholera, plague, malaria, and typhus. He won the 1905 Nobel Prize in physiology and medicine for his work toward the germ theory of disease.

Questions to Review

What did Robert Koch find in his studies of anthrax?

Why is anthrax a potentially dangerous biological weapon?

2. Technology Connection: Throat Cultures

You have probably had strep throat. It is caused by a variety of the same bacteria that cause childbed fever. Strep throat is not especially dangerous, but in rare cases it can develop into rheumatic fever, which is quite serious. Therefore, when you have a very sore throat, you will often be tested for strep throat. Because strep throat is highly contagious, it's important to detect and treat it promptly to keep it from spreading.

Questions to Review

How does the rapid strep test work?

Why would doctors bother to do the culture strep test that takes longer, when they get quick results from the shorter test?

If strep throat isn't all that dangerous, why is it important to test for it at all?

2. Life Application: Childbed Fever

Childbed fever was very common in cities in Europe and America between 1700 and 1900. The connection between the disease and hospitals was clear, and there was evidence of the link between doctors and the disease. In the 1840s, Oliver Wendell Holmes, a physician in Boston, published a paper arguing that doctors needed to wash their hands and sterilize their instruments in order to reduce transmission of the disease. It was not until the early 1900s that a majority finally accepted that childbed fever and other contagious diseases were caused by germs and transmitted unless attention was paid to hygiene and sterilization.

Questions to Review

If there was solid evidence that childbed fever transmission was connected to hospitals and doctors as of the 1840s, why did it take another half century for proper hygiene and sterilization techniques to be widely implemented?

How are we now able to treat bacterial infections like this such that they don't often kill people anymore?

2. How Do We Kknow?: Hypothesis Testing and Scientific Proof

Science is based on inductive reasoning, where conclusions follow from evidence. For my hypothesis to be correct, *if* I set up particular conditions, *then* this should happen. If it does happen, my hypothesis is supported, but it is not *proven*. It is still possible that evidence could come up that would prove it false, or I might not be considering something. The hypotheses are *falsifiable*, or make predictions that could fail. If the hypothesis is tested repeatedly in different ways and the predictions continue to hold, a scientist can conclude the hypothesis is essentially proven until contradictory evidence arises.

Questions to Review

Hypotheses must be falsifiable. What does that mean and why does it matter?

Why aren't hypotheses ever considered "proven" in science?

Student Study Guide

Chapter 3 Human Development

How Do Cells Make A Person?

This chapter illustrates the developmental path from a single cell to a complete human being. The journey depends on cell communication, cell movement, and the ability to regulate how genetic information is used.

3. Case Study: Unusually Close Sisters

3. Case Study Summary

Abigail and Brittany Hansel were born in 1990 as conjoined twins. Each has her own head, neck, nervous system, heart, esophagus, and gall bladder but they are joined sided to side and share many other organs. Their spines are joined at the pelvis so they have just two arms and two legs. A model of cooperation, they walk, play piano, swim, ride horses, play basketball, and type. Both have driver's licenses. But they have quite different personalities, likes, dislikes, food preferences, and are good at different subjects in school. They share the same genes and are, literally, stuck together. How can they be so different?

Question for Review

Explain what a conjoined twin is.

Learning Objective 1: What are the Units of Life?

   Identify the most basic units of life.

3.1   Summary

All living things develop from and are made of cells, the basic unit of life. The process of development from cell to complete multicellular organism is embryogenesis, and scientists who study it are called embryologists.

Before microscopes were invented, cells could not be seen, so scientists thought that a healthy body required four humours: blood, phlegm, yellow bile, and black bile. Illness occurred when these were not in balance, so a common treatment was bloodletting or the removal of blood to restore balance.

Autopsies determined that organs and blood circulation existed, along with tissues and fibers, and eventually, cells were identified. Once they could be seen, it was found that every living thing was made of cells, and cells only came from other cells: cell theory.

3.1 Define these Key Terms

embryology

embryogenesis

cell theory

3.1 Questions for Review

3.1.1 Explain how early scientists thought a healthy body worked.

3.1.2 How did scientists determine that organs, blood circulation, and tendons and fibers existed, and how did they eventually find cells?

Learning Objective 3.2: What Cell Structures Play a Role in Embryo Development?

Describe the cell structures that play a role in embryo development

3.2 Summary

Humans are made of eukaryotic cells. These are just like eukaryotic cells in other organisms, evidence that all species share a common ancestor. Eukaryotic cells include the following:

Cell Membrane

The cell membrane is the outer boundary of the cell that separates it from the environment and regulates what goes in and out. It encloses the cytoplasm, which includes the viscous cytosol where molecules are dissolved, and the organelles. It can detect chemical signals from other cells, and it can stick cells together.

Nucleus

The nucleus is the most prominent organelle in the cell and houses the chromosomes, which contain the genetic information of the cell in the form of genes made of DNA. Gene expression, or when genes are turned on and off, controls what types of cells form and their location in the developing embryo.

Mitochondria

This organelle extracts energy from food to power the cell. High-energy cells, like muscle cells or developing embryonic cells, have many mitochondria.

Endomembrane System

The endomembrane system produces molecules and delivers them to key locations within and outside the cell. It also collects, packages, and removes waste materials. It includes three parts:

- o Endoplasmic reticulum (ER): This is a series of membrane folds near the nucleus. The smooth endoplasmic reticulum (SER) handles the synthesis of fats and lipids, and breaks down and eliminates some toxins. The rough endoplasmic reticulum (RER) produces proteins for cell structure and regulation. Ribosomes stud its surface and thread the proteins into the RER, where they are packaged in vesicles to travel to other parts of the cell.
- o Golgi apparatus: Here, proteins acquire their final structure and, in some cases, travel in vesicles to the cell membrane for release outside the cell.
- o Lysosomes: These organelles digest waste materials and worn out organelles and recycle molecules.

Cytoskeleton

The cytoskeleton maintains the shape of the cell, the positions of its organelles, and allows for cell motion. There are three main parts:

- o Microtubules: These determine cell shape and make up flagella which are "tails" on some cells (like sperm cells).
- o Microfilaments: These also aid in cell movement and pinch together the cell membrane so a cell can divide in two.
- o Intermediate filaments strengthen cells.

3.2 Define these Key Terms

cell membrane

nucleus

mitochondria

endoplasmic reticulum

golgi apparatus

lysosomes

cytoskeleton

Learning Objective 3.3: How Do Eggs and Sperm Form?

Explain how eggs and sperm form through meiosis.

3.3 Summary

Gametes, or sex cells, are produced by all sexually reproducing species. Males produce sperm cells and females produce eggs by a process called meiosis. Human cells have 23 pairs of chromosomes, and because the gametes combine to create a new human, they must have half the pair or else the new human's cells would have 46 pairs of chromosomes. Meiosis produces four haploid gametes (each gamete has half the chromosomes) from a normal diploid cell (one with a complete set of chromosomes).

Sperm cells are produced by the male testes and egg cells are produced by the female's ovaries. The process of meiosis involves two sequences:

Meiosis I:

- Interphase: Chromosomes replicate and are loosely packaged.
- Prophase I: Chromosomes become short and thick and line up with their pairs. The nuclear membrane comes apart, leaving the chromosomes in the cytoplasm.
  - The spindle is formed, with a group of microtubules extending from each end of the cell that will snag the chromosomes and drag them to separate sides of the cell.
- Metaphase I: Paired chromosomes line up single file down the middle of the cell.
- Anaphase I: Paired chromosomes separate toward opposite poles of the cell.
- Telophase I and cytokinesis: The cell membrane pinches together and divides in two, with each new cell containing a complete set of chromosomes and organelles. Now the two copies of the chromosomes in each cell, the chromatids, must be separated.

Meiosis II: This proceeds in both new cells.

- Prophase II: The chromosomes attach to a newly formed spindle.
- Metaphase II: The chromosome pairs attach to the spindle and line up down the middle of the cells.
- Anaphase II: The chromatids of each chromosome detach and move to opposite ends of the cell.
- Telophase II and cytokinesis: The nuclear membrane re-forms around the chromosomes and cytokinesis pinches the cells in two.

At the end of this there are four haploid cells or gametes.

3.3 Define these Key Terms

gametes

meiosis

testes

ovaries

haploid/diploid cells

3.3 Questions for Review

3.3.1. Without looking at the diagram now, draw your own diagram of cells going through meiosis I and II. Note how you end up with four haploid cells at the end.

3.3.2. Explain why it matters that the gametes are haploid cells.

Learning Objective 3.4: What Happens in Fertilization?

Summarize what happens in fertilization.

3.4 Summary

In most mammals, like humans, sperm swim to the egg inside the body of the female. The human egg has a protective shield, the *zona pellucida*, that has proteins that only bind with one male human sperm cell. Once the one sperm and one egg have fused, the resulting zygote, or one-celled embryo, will have a complete set of chromosomes.

3.4 Define this Key Term

zygote

3.4 Questions for Review

3.4.1 How does a human egg allow for only one <u>human</u> sperm to fertilize it?

3.4.2 What is a zygote and is it haploid or diploid?

Learning Objective 3.5: How Does an Embryo Form and Ultimately Become a Fetus?

Outline the stages of mitosis and the ways that cells become specialized.

3.5 Summary

Embryogenesis begins with cleavage; the original fertilized egg replicates and divides to form a multi-celled embryo and eventually a complete human. This type of cell division is called mitosis. This time, all of the chromosomes are copied, and each new cell gets a complete copy, identical to the original cell. It remains diploid, so the process is a bit simpler than meiosis.

Mitosis:

- Interphase: The chromosomes replicate and the cell gets bigger and makes the materials it needs to divide.
- Prophase: The chromosomes become shorter and wider, the nuclear membrane disappears, and the spindle forms.
- Metaphase: The chromosomes line up single file down the middle of the cell attached to the spindle.
- Anaphase: The chromatid pairs of chromosomes separate such that one full set goes to each pole of the cell.

- Telophase and cytokinesis: A nucleus reforms around each set of chromosomes and the cell pinches into two cells, each with a complete set of chromosomes identical to the parent cell.

In mammals, like humans, when the embryo reaches the 16-cell stage, the outer cells will produce the placenta, the organ that connects the fetus to the mother's uterine wall, as well as other structures necessary to support the fetus. The inner cells are the embryonic stem cells and can and will produce all the types of cells that make up a fully developed human.

Gastrulation and Organ Formation

During gastrulation, cells and tissues move to locations, where they will grow into organs. For example, the central nervous system starts out as a flat sheet of cells in a tissue that rolls up to form the neural tube. The front end of the tube expands to become the brain and the back end becomes the spinal cord. Any mistakes here have serious consequences. For example, if the front end of the neural tube doesn't close, the fetus will have no brain.

Differentiation

The zygote can produce any type of cell in the body, and an adult human has more than 200 specialized cells. That change, from a cell that can do anything to a more specialized cell, is called differentiation. The cells become increasingly specialized as the embryo develops. During gastrulation, three major types of embryonic cells emerge: ectoderm, mesoderm, and endoderm. Muscle cells come only from the mesoderm, and some of those will be cardiac (i.e., heart muscle cells). Once they move to the heart and differentiate into those, they can't do anything else.

Gene Expression

Cell differentiation depends on communication between cells and on gene expression. All cells contain the same genes, but different genes are used in different cells.

3.5 Define these Key Terms

embryo

cleavage

mitosis

cell differentiation

gene expression

3.5 Questions for Review

3.5.1 Without looking at the diagram of mitosis, create your own. What is the end product of mitosis *versus* meiosis?

3.5.2. Explain what gastrulation is.

3.5.3 What is the difference between an embryonic stem cell and a differentiated cell?

3.5.4. What is meant by gene expression, and how does it related to what a cell does in the body?

Learning Objective 3.6: What are the Key Events of Pregnancy?

Enumerate the key events in pregnancy.

3.6 Summary

The odds of conceiving during a single fertility cycle are just 15–25%, depending on age, health, and other factors. Then a series of steps must all work out right to produce a healthy newborn.

The Stages of Pregnancy

1. Embryonic Development: The First 8 Weeks

- Week 1: Cleavage produces cells that will form the embryo and chorion, which is part of the placenta.
- Week 2: The embryo, about the size of the period at the end of this sentence, attaches to the uterine wall, or womb. Gastrulation produces the ectoderm, mesoderm, and endoderm.
- Week 3: Neurulation, or formulation of the central nervous system begins, along with heart development.
- Week 4: Limb buds form.
- Week 5: Nose, eyes, and ears appear.
- Weeks 6 and 7: Fingers and toes, and a skeleton of cartilage.
- Week 8: All the organ systems are developing, and the embryo is 1.5 inches long, and fetal development begins.

2. Fetal Development: 3 to 9 Months

- Month 3: Gender can be discerned and fingernails form; the placenta is fully developed.
- Month 4: Fetus is approximately 6 inches long and weighs 6 ounces; the skeleton is visible in ultrasound imaging, hair is present, and an experienced mother can feel fetal movement.
- Month 5: Fetal heartbeat can be detected with a stethoscope.
- Month 6: Organ systems continue to mature.
- Month 7: Fetus is 12 inches long, weighs 3 pounds, has open eyes, and is covered with a fine hair called lanugo. The fetus is now oriented head downward, and if it were born now, it would have a 90% chance of survival.
- Months 8 and 9: Growth and maturation of the lungs, disappearance of body hair, and deposits of fat for insulation. The typical newborn is about 20 inches long and weighs 7 pounds.

3.6 Questions for Review

3.6.1 Describe a few of the events which occur in embryological development. When does this end?

3.6.2 Describe an embryo at the end of embryological development.

3.6.3. Describe a few of the events which occur in fetal development. When is this period in a pregnancy?

Learning Objective 3.7: What Happens in Labor and Delivery?

Describe the stages of labor and delivery.

3.7 Summary

Normally, when the fetus is ready to be born, it secretes surfactant, which helps prevent the lung cells from sticking together and interfering with breathing. This causes an immune response in certain fetal cells which then migrate to the uterus, where they secrete a chemical signal telling the uterus to contract. Labor and delivery can be divided into three stages:

1. Uterine contractions pull the lower part of the uterus up toward the fetus's head, and push the fetus down. This pressure on the cervix, or the lower end of the uterus, causes it to stretch or dilate. The contractions also release the hormone oxytocin, which stimulates more contractions. The amnion, or sac in which the embryo is suspended, breaks, releasing amniotic fluid. The cervix dilates to about 4 inches.

2. Contractions are very strong, occurring every minute or so, the mother has an irresistible urge to push, and the baby is born.

3. The placenta is delivered.

Humans are born less well developed than many other mammals, possibly because of the size of the human head. If the fetus got much larger, delivery would be impossible without surgery, as is sometimes the case, resulting in *Caesarian* deliveries.

3.7 Define these Key Terms

cervix

amnion

3.7 Questions for Review

3.7.1 Draw a flowchart outlining the steps of labor and delivery.

Learning Objective 3.8: How Do Twins Form?

Explain how twins form

3.8 Summary

Twins are born in a little more than 1% of pregnancies; 3% with fertility treatments.

Identical twins come from a single fertilized egg that splits into two embryos early in development. There are three categories of identical twins:

- In 33%, the embryos have completely separate chorions, which means they separated within 5 days of conception.
- In 66%, the embryos share a chorion but have separate amnions (the fluid filled sac that surrounds the embryo). This means that the embryos split between 5 and 9 days post-conception.
- About 1% of identical twins share both a chorion and amnion. This means separation occurred later than 9 days after conception. These twins are at risk of being conjoined.

Fraternal twins result when the mother released two eggs, and each was fertilized by a separate sperm. They are no more closely related than any siblings born to the same parents, and they need not be the same gender.

3.8 Questions for Review

3.8.1 Explain the difference between identical and fraternal twins.

3.8.2. At what point do identical twins have to separate in the womb to be at low risk of being conjoined?

Learning Objective 3.9: What can Conjoined Twins Tell Us About Biology and Ourselves?

Assess what conjoined twins can tell us about biology and ourselves

3.9 Summary

Forty to sixty percent of conceptions result in miscarriages, where the embryo does not survive. Usually this is because of genetic or physical defects. Conjoined twins form in 1 of every 50,000 pregnancies, but they result in only 1 in 200,000 births, and more than 75% of those will not survive.

Side-by-side conjoined twins like Abigail and Brittany are very rare. They can also be joined at the buttocks, the hip, or head to head, and the angle of joining varies, as does the size of the twins. There is no clear explanation for why this happens. It's more common in females and occurs more frequently in India and Africa, less in China and the United States. Some scientists argue that they didn't separate properly as embryos, while others argue that they "stick" together at some point in development.

Conjoined Twins Tell Us About the Biology of Development

Abnormal biology often teaches us more than studying "normal" events. Conjoined twins help us understand the mechanisms by which an embryo forms a body. For example, it's not unique to humans, and it can be induced in other organisms. Experiments like this have taught scientists about the molecular signals, cell interactions, and gene expression that happen in both conjoined and normal development.

The fact that twinning can occur at least up until gastrulation at 14 days makes it hard to argue that we are defined as individuals before this time, so it's not clear when "personhood" is established, even biologically.

Conjoined Twins Tell Us About Ourselves

In ancient times, conjoined twins might be worshipped as gods, banished, or killed. They often have been treated as side-show freaks, displayed for entertainment. Chang and Eng Bunker were the original "Siamese Twins" and had lucrative careers in show business. Now, separation is generally attempted, but sometimes the twins share a vital organ and cannot be split.

Finally, the fact that Abigail and Brittany, and Cheng and Eng, too, have such different personalities shows us that our genes do not solely determine who we are.

3.9 Questions for Review

3.9.1 Scientists can induce several organisms to produce conjoined offspring. Why would they do this? What can they learn?

3.9.2   What do conjoined twins teach us about genetics and personality?

3. Biology in Perspective

A haploid egg and sperm cell combine at fertilization to produce a diploid zygote, and those genetic instructions, from both the father and mother, direct mitosis, cell movement, communication, adhesion and differentiation, and gene expression to produce a complete human. Each human is unique, even if they share genes like identical twins do. Humans are highly variable such that it's hard to define "normal." Question to consider: What does it mean to say that a human is more than just biology?

3. Scientist Spotlight: Anton van Leewenhoek (1632–1723)

In 1668, Anton van Leeuwenhoek, a fabric merchant, learned how to grind lenses and make simple microscopes. Just out of curiosity, he examined various things and found bacteria in his saliva, protozoa, tiny invertebrates, and algae in pond water, and red blood cells in his blood. He found sperm in his own semen and also in the semen of other animals. He also found the sperm in the uterus and Fallopian tubes of recently mated dogs and rabbits. His observations were the beginning of our understanding of the cellular basis of reproduction.

He lacked any formal education, but the value of his observations got him elected into the Royal Society of London, where he sent his detailed descriptions. Question to consider: What role does curiosity play in scientific discovery?

3. Technology Connection: Ultrasound

These days, the first picture of a baby is likely to be the fuzzy image of a tiny fetus: an ultrasound image. Ultrasound "sees" with sound waves, like a bat or a fish finder on a fishing boat. High-frequency sound waves echo back from an object and are sent through a recording device called a transducer and are transformed into a digital image in real time. During pregnancy, ultrasound is used to monitor fetal growth, development, and the health of the placenta, and it can identify developmental problems.

Ultrasound is also used to examine major organs and blood vessels whenever there is a reason for a doctor to look at these. It can also be used to guide biopsies (i.e., tissue sampling).

Questions to consider:

How does ultrasound work, and what is it used for in modern medicine?

3. Life Application: Fetal Alcohol Syndrome

It has been determined that a pregnant woman drinking alcohol is dangerous for the developing fetus. It can result is Fetal Alcohol Syndrome (FAS). This can cause poor growth, abnormal facial features, hyperactivity, poor reasoning skills, vision problems, hearing problems, and mental disabilities. These children are at higher risk for psychiatric problems and criminal behavior. There is no treatment, although a supportive home and education can help. The ailment is completely preventable if the pregnant woman avoids alcohol, but it often takes at least a few days or longer for a woman to realize she is pregnant.

Question to consider:

What is fetal alcohol syndrome and how can it be prevented?

3. How Do We Know: Eggs and Sperm are both Needed for Fertilization

Anton van Leeuwenhoek discovered sperm cells in semen in 1667, and he proposed that, in fertilization, sperm enter the egg, but mammalian eggs were not discovered until 1797. In 1784, an Italian scientist put little trousers on male frogs, allowed them to mate, and found that they were unable to produce offspring. He also found, when he combined frog sperm with eggs directly, that fertilization took place. In 1824, French scientists determined that sperm cells, not just semen, were necessary to trigger development in amphibian eggs. In 1854, a British scientist finally watched frog sperm enter frog eggs. In 1976, a German scientist demonstrated that, when egg and sperm from a sea urchin fuse, the nuclei fuse as well, combining the genetic material from each parent.

Question to consider:

List the sequence of events from discovery of sperm cells to the understanding of how offspring contain genetic material from each parent.

Student Study Guide

Chapter 4: Inheritance, Genes, and Physical Characteristics

4. Case Study: Does disease have a genetic basis?

4. Case Study Summary

Malaria is a disease caused by a protozoan transmitted to people by a mosquito bite. As a result of contracting this disease as a child while visiting Africa, Tony Allison decided to become a physician. He was also interested in human populations. On a later return trip to Africa, he collected blood samples from people from various tribes. He looked at the proteins in their red blood cells in hope of identifying patterns of genetic relationships among the tribes. He also tested for sickle cells, abnormal red blood cells indicative of a painful, and at that time, generally fatal disease. Allison knew that the density of mosquitoes varied from place to place in Africa, and he found that where mosquitoes were abundant, malaria was prevalent and sickle cells were common. Where it was dry and mosquitoes were uncommon, malaria and sickle cell were both rare. Allison hypothesized that having the sickle cell trait must provide some resistance to malaria.

In 1952, Dr. Allison returned to Africa with the goal of testing his hypothesis. He designed three studies:

1. Scientists injected the malarial parasite into healthy volunteers. When malaria developed, the scientists administered a test drug to see whether it was effective. Allison took blood samples from all 30 of these test subjects, and he found that only 1 of the 15 subjects who had the sickle cell trait had developed malaria. Fourteen of the 15 people who did not have sickle cell did develop malaria.
2. Maybe those people with resistance simply had been previously exposed and their immune system was able to resist the malaria? This time he looked for the malaria protozoan in blood samples of 290 children. Only 28% of the children who had the sickle cell trait had the parasite. And 46% of the children who did not have the sickle cell trait were infected.
3. Finally, Allison took blood samples from 5000 individuals representing more than 30 African tribes. In all cases, sickle cell was rare or absent where malaria was absent and common where malaria was a serious problem.

These studies supported Allison's hypothesis that the sickle cell trait confers resistance to malaria. The rest of the chapter looks at how these two things can be related.

Question for Review

Explain how Dr. Allison first saw the connection between mosquitoes and malaria, and then how he found the connection between malaria and sickle cell disease.

Learning Objective 1: What is sickle cell disease?

Specify the causes and consequences of sickle cell disease and the relationship between sickle cell disease and malaria

4.1   Summary

In 1910, red blood cells shaped like commas or sickles were observed in a student suffering from anemia. Symptoms of anemia include fatigue, dizziness, shortness of breath, and chest pain. In 1917, it was observed that in a person with sickle cell disease, the red blood cells sickle (become sickle-shaped) as they are deprived of oxygen.

Normal red blood cells are disk shaped and flexible, but sickle cells are curved, pointy, and rigid. They tend to get stuck in narrow blood vessels, forming clots which block blood flow to the cells, tissues, and organs "downstream" from the clot, thus depriving them of the oxygen they need for energy. This can cause muscle cramps, intense pain, anemia, swollen joints, kidney failure, and heart failure. Sickle cell disease is the most commonly inherited blood disorder in the United States and is particularly prevalent among African Americans.

The problem is related to abnormal hemoglobin. Normal hemoglobin (HbA) is the protein that carries oxygen molecules in the red blood cells in the blood to deliver them to the cells and tissues of the body. Sickle cell hemoglobin (HbS) can carry oxygen, but when it gives up its oxygen, it forms long fibers that cause the red blood cells to become sickle shaped and potentially clot.

If a person inherits the genes that encode for the HbA protein, he or she will not have sickle cell disease. If he or she inherits one normal gene and one that produces the HbS protein, the person will be a carrier of the disorder but will not develop the full-blown disease. If an individual inherits two HbS genes, the person will have sickle cell disease. The carrier and the person with the disease are both resistant to malaria. There currently is no cure for sickle cell disease, and at least 8% of African Americans are carriers.

4.1 Define these Key Terms

malaria

protozoan

red blood cell

anemia

hemoglobin

HbA

HbS

4.1 Questions for Review

4.1.1 Sickle cell disease can be deadly, yet it's fairly common among people who come from certain areas of the world. Why is this?

4.1.2 Review Dr. Allison's three early experiments with malaria and sickle cell. What was the set up of each and how did the results advance Dr. Allison's idea that sickle cell and malaria resistance were somehow related?

Question addressed	Experiment	Results	Conclusions

4.1.3 Sickle cell disease is an inherited trait brought on by the gene involved in producing hemoglobin. What does this gene do that can cause sickle cell? If you have one of these genes, do you necessarily have sickle cell disease? Why or why not?

Learning Objective 4.2: Could molecular medicine prevent sickle cell disease?

Explain how Linus Pauling's research into sickle cell disease ushered in the era of molecular medicine.

4.2 Summary

Linus Pauling is one of the most important chemists of the 20th century, in part because he was the first person to consider diseases at the molecular level—the start of molecular medicine. He happened to be at a conference at Harvard University in 1945 where he learned about sickle cell disease, and he wanted to figure out what was different about the abnormal hemoglobin protein. Using the technique **electrophoresis** (see Technology Connection: Electrophoresis summary below) he determined that HbA and HbS were different chemically, carriers contained equal amounts of HbA and HbS, and HbS forms filaments at low oxygen levels and this causes sickling. He thus figured out that the hemoglobin present in a person was related to the genetics of the disease, that the disease was caused by an alteration to the molecular structure of the hemoglobin protein, and therefore that genes must determine protein structure.

The technology was not available to study the precise difference between HbA and HbS in Pauling's time, but by 1957 it was known that proteins were made of chains of amino acids, and it was discovered that HbS had one wrong amino acid in the chain. This one change produces HbS rather than HbA and results in the sickle cell trait.

4.2 Define these Key Terms

electrophoresis

molecular medicine

amino acid

4.2 Questions for Review

4.2.1 What is the job of hemoglobin in the body? Given this, what exactly is the problem in sickle cell disease?

4.2.2 How did Linus Pauling determine that HbA was different from HbS.

4.2.3 What are proteins made of?

Learning Objective 4.3: Where is our genetic information stored?

Describe the study that disproved the concept of pangenesis

4.3 Summary

Humans have been interested in heredity since farmers first realized they could selectively breed plants and animals for particular traits. If you breed a particularly fast horse with another particularly fast horse, at least some of those offspring will likely also be fast. For a long time people thought this was the result of pangenesis. The idea was that each part of the body contained a characteristic "seed" which traveled to the reproductive organs and when mating occurred, the "seeds" from the various parts of the two parents combined to produce offspring with similar characteristics. In the 1880s, August Weismann tested this explanation. He took a population of mice and removed their tails, thus presumably eliminating tail "seeds." This ought to result in offspring with short or no tails, right? He did this for 23 generations and concluded that the information necessary to produce a tail didn't actually reside in the tail itself. Pangenesis isn't the explanation for heredity.

4.3 Define this Key Term

pangenesis

4.3 Questions for Review

4.3.1 Pangenesis as an explanation of heredity was around for a long time, before we understood genes. How was it supposed to work and why do you think it was accepted so easily for so long?

4.3.2 August Weismann eventually demonstrated that pangenesis did not work to explain heredity. (*Diagram here to fill in blanks/explain.*)

   a. What was his experiment?

   b. If pangenesis were correct, what would have happened?

   c. What did happen?

Learning Objective 4.4: How did Mendel discover the rules of inheritance?

Outline how Mendel conducted his experiments related to inheritance and the resulting set of rules.

4.4 Summary

Twenty years prior to Weismann's work with mice, an Augustinian monk named Gregor Medel was working on the question of inheritance as well. He worked with pea plants. He looked at one trait or characteristic at a time (yellow *vs.* green peas, white *vs.* purple flowers) in offspring resulting from crosses (matings) of parents with known traits. He followed these traits over generations and he counted the number of offspring in each generation with each trait. Look at the example in your book (Figure 4.14) in which he examined a cross between parent plants with purple flowers and plants with white flowers. The first generation of offspring all produced purple flowers, but in the second generation (the grandchildren), white flowers show up again in about one quarter of the offspring. Mendel figured that flower color must be regulated by two hereditary factors, one from each parent, and in this case, the purple factor must be dominant and mask the white factor in the first generation, but those offspring must be carrying the white factor for it to show up in the second generation.

We now know that these "factors" are genes which sort into different gametes in the parents and so get passed on to the offspring. Mendel didn't know this, but he was a very good statistician and he knew that if you had two independent factors sorting into pairs you'd get particular ratios in the offspring. In the case of the flower color, you'd get a 3:1 ratio of purple to white in the grandchildren. And that's just about what he got.

This type of cross where we're just looking at one trait is called a monohybrid cross. Mendel analyzed six more types of monohybrid crosses in the pea plants and in every case got that 3:1 ratio in the second generation. He didn't understand the mechanism, but he figured out that each trait is encoded by two factors, one from each parent.

Sickle cell disease follows this pattern. A person with sickle cell disease has two genes, both of which code for HbS. Someone with normal hemoglobin has two genes which both code for HbA. And a carrier has two genes, one coding for HbA and one coding for HbS.

Mendel's Rules

1. Genes can come in more than one form or allele. So, for example, they can code for white or purple flowers, or HbS or HbA proteins)

2. Alleles of a particular gene sort individually into games during meiosis. So, for example, a pea plant with a purple allele and a white allele would produce 50% purple gametes and 50% white gametes.

3. Some alleles are dominant and some are recessive. So, for example, if you have a pea plant with both a purple and white allele, since purple is dominant, only purple will show up. The only way a pea plant can have white flowers is if it carries two copies of the recessive white allele.

4.4 Define these Key Terms

trait

cross

dominant

recessive

gamete

gene

allele

monohybrid

4.4 Questions for Review

4.4.1 Your book illustrates Mendel's experiments examining the heredity of purple and white color in pea plants. Diagram the same experiment with two generations of crosses (go through to the "grandchildren") but looking at seed color instead of flower color. What would be the "factors" (we now know they're alleles) present in the offspring and in what ratios?

4.4 .2 Repeat the exercise above but start with the following parents: normal HbA X sickle cell HbS

What percentage of the offspring in the second generation have full blown sickle cell disease? What's the percentage of this in the first generation? Why?

4.4.3 How does a trait present in a parent "skip a generation," i.e. not show up in the children but show up in the grandchildren?

4.4.4 Explain how genes and alleles are related to each other.

Learning Objective 4.5: How much do Mendel's rules explain?

Identify the three reasons why Mendel's rules fail to explain inheritance completely.

4.5 Summary

Now we know that Mendel figured out the basics of heredity but it's a bit more complicated than he found with his peas. There are three main reasons for this:

1. Alleles can interact with each other, as can genes.

   Sickle cell is actually a good example of this. Neither the normal HbA nor the abnormal HbS is dominant over the other. So a person with both an HbA and an HbS allele has a mild version of sickle cell disease.

Genes can interact like this too. You might have hair color alleles in the right combination to give you red hair, but if you also have the no-hair allele for baldness, you won't have any hair, red or no. In this case the baldness gene would override the hair color genes.

2. Genes may affect more than one characteristic.

    Most genes affect more than one characteristic. Again, the sickle cell gene is a good example. It can give you sickle cell disease, but it can also make you less susceptible to malaria. The genetic disease cystic fibrosis works this way too. The CF allele causes a serious disease with no cure yet, but it also confers some resistance to cholera and other infectious intestinal diseases

3. Gene expression depends on the environment.

    Sometimes the trait a gene produces depends on the environment in which it's working. The flowers of a *Hydrangea* plant range from blue to purple to pink depending on the acidity of the soil in which they grow. This is an especially important consideration during fetal development when environmental characteristics can have an especially strong influence.

4.5 Define this Key Term

cystic fibrosis

4.5 Questions for Review

4.5.1 The disease cystic fibrosis (CF) is genetic like sickle cell disease. In CF the lungs and digestive system accumulates mucous and without careful medical intervention, people with two CF alleles aren't likely to live very long. However, carriers have been found to be resistant to cholera. What's going on genetically and how is it similar to the situation with sickle cell carriers?

4.5.2 These two hydrangea plants are identical twins even though one has pink flowers and one has blue flowers. (*Picture from book?*) Their genes are the same. What's going on?

Learning Objective 4.6: What are genes made of?

Detail the structure of DNA.

4.6 Summary

Genes are made of DNA (deoxyribonucleic acid) which is a type of nucleic acid, called that because it's found in the nucleus of the cell. It is DNA that holds our genetic information.

DNA is composed of four different nucleotides: adenine (A), thymidine (T), guanine (G), and cytosine (C). These are connected to each other by a sugar-phosphate backbone with the nucleotides sticking out off one side. When these nucleotides pair up with each other, A to T and C to G (they only pair like that such that a chain of AATCG will only ever pair up with a complementary chain TTAGC), you get the ladder-like structure of DNA which twists to form the double helix shape characteristic of a DNA molecule.

4.6 Define these Key Terms

DNA

nucleotides

complementary base

4.6 Questions for Review

4.6.1 (also "How do we know?") Assume that dark hair and blonde hair alleles are dominant and recessive like Mendel's peas—no other interactions except that a <u>carrier</u> of blonde will have hair color lighter than dark, but darker than blonde, i.e., be a mix of the two colors. Draw a pedigree for your family back through your grandparents if you know their hair colors.

Learning Objective 4.7: How does DNA function?

Describe how scientists uncovered the physical basis of genes.

4.7 Summary

There were two initial steps to determine how DNA works:

1. Frederick Griffin figured out that a substance in bacteria could transform how other bacteria work.

    Griffin had a type of bacteria that existed in two forms. One was smooth and caused pneumonia, and the other one was rough and harmless. He wondered what made the difference so he manipulated the bacteria in numerous ways and then injected it into mice to see if it would still cause pneumonia. In one case, he killed the smooth bacteria and combined these dead bacteria with living rough bacteria. He infected the mice with the dead smooth bacteria, the rough bacteria, or the combination of the two. Only the mice with the combination got pneumonia, and when he looked at the bacteria infecting those mice, it was all smooth. The formerly rough bacteria had been **transformed** by the smooth, and since this trait persisted over generations, the change was heritable. He knew this was a big deal, but he didn't know what the "transforming substance" was. World War II then interrupted progress on this.

2. Oswald Avery, Colin MacLeod, and Maclyn McCarty showed that the substance doing the transforming was DNA.

    In 1944, this group of scientists performed a series of experiments similar to Griffith's except they isolated the various components of a cell that could possibly be the transforming substance and looked at what happened one substance at a time: carbohydrates, proteins, fat, RNA, and DNA. Only the mixture containing DNA combined with rough bacteria resulted in smooth bacteria. DNA was the transforming substance.

4.7 Define this Key Term

transformation

4.7 Questions for Review

4.7.1 Diagram Frederick Griffin's experiment with the smooth and rough bacteria in mice.

4.7.2 Diagram the experiment that showed that DNA was the "transforming substance" in the bacteria above.

Learning Objective 4.8: What processes must DNA accomplish?

Diagram the processes of DNA replication, mutation, and protein production.

4.8 Summary

DNA has to do three things:

1. Replicate itself
2. Be able to change or **mutate**
3. Produce proteins responsible for physical changes

1. Replication

When cells divide during mitosis, the DNA must replicate so the new cell has a complete copy of the original DNA. Likewise in meiosis, the chromosomes (which are made of genes which are made of DNA) must replicate. James Watson and Francis Crick discovered the double-helix structure of DNA, which neatly suggested how replication could work. The molecule can "unzip" and each strand (each half of the ladder) then recreates a "complementary strand" based on the fact that the nucleotide A will only pair with T and C will only pair with G.

2. Mutation

A mutation is simply a change in DNA. If DNA never mutated, genes would remain identical from one generation to the next and so living things would never change. Given the huge diversity of life on Earth, obviously DNA, and therefore living things, can change. This can happen because of an error during replication. DNA replication is amazingly accurate, but it isn't perfect. Or DNA can be damaged. There are fives types of mutations:

1. Point mutation: This is like the case with the sickle cell HbS. One nucleotide was replaced by a different one and that one mutation resulted in the abnormal HbS.
2. Deletion: In this case a nucleotide or segment of DNA is removed.
3. Duplication: A segment of DNA gets repeated within a chromosome.
4. Inversion: A segment of DNA gets reversed within a chromosome.
5. Translocation: A segment of DNA moves from one chromosome to another.

In most cases, there is no physical effect of the mutations. Sometimes the mutation is lethal, either before the organism develops or afterward, as in certain rare cancer-causing genes. In some cases the change produces an allele that is beneficial in some way and gives a selective advantage to those who possess it. For example, a mutation to the human gene CCR5 prevents binding of the human immunodeficiency virus (HIV) to immune cells and so those who possess this allele are resistant to AIDS.

Sometimes mutations are the result of external influences such as ultraviolet light (from sun, tanning beds), radiation, certain industrial chemicals, and activities like smoking tobacco. In some cases, the cell recognizes and repairs these, but not always.

3. Protein Production

The job of DNA is to direct synthesis of the proteins that make a body work. There are three parts to this process: the genetic code, transcription into RNA, and translation into proteins.

1. Genetic Code
The alphabet of the genetic code consists of the four letters A,T,C and G. The letters are read as words that are three letters long and are called **codons.** Each three-letter codon corresponds to a specific amino acid that will be incorporated into a protein. There are a total of 64 codons which code for the 20 different amino acids necessary to make the proteins that make up an organism (most of the amino acids are coded for by more than one three-letter codon).

2. Transcription
DNA is translated into RNA which is similar to DNA except
   a. The sugar in the nucleotides is ribose, not deoxyribose.
   b. The nucleotide uracil (U) is used instead of T, and U pairs with A.
   c. RNA is single stranded, not a double helix like DNA.

The DNA in the nucleus unwinds and the nucleotides on each strand are paired with nucleotides to make a **messenger RNA (mRNA)** strand that then detaches and moves into the cytoplasm. **Transfer RNA (tRNA)** pairs up **anticodons** based on each codon on the mRNA. An anticodon is the complementary three-letter strand of RNA. On the other end of the tRNA molecule from the anticodon is the amino acid coded for by the anticodon. Finally, **ribosomal RNA (rRNA)** forms ribosomes, which hold the mRNA and tRNA together such that the amino acids can be connected in the correct order. Those amino acids will make a protein.

4.8 Define these Key Terms

replication

mutation

genetic code

transcription

translation

messenger RNA

transfer RNA

ribosomal RNA

anticodon

codon

4.8 Questions for Review

4.8.1 Here is half of a DNA strand. Fill in the complementary strand.

4.8 2 Here is that same half of DNA. Now fill in a complementary strand with a mutation. What are the possible consequences of this? Is it always bad?

Learning Objective 4.9: Why is protein structure so important?

Explain why a protein's structure is so important to its function.

4.9 Summary

Amino acids come in all different sizes and shapes and have differing electrical charges. A particular string of amino acids will therefore fold up together into a particular shape. If any amino acid is changed or omitted (i.e., there is a mutation), the protein will not fold correctly and that change in shape will likely dramatically change its function. The shape of a protein is what matters.

Look at hemoglobin. The shape of the HbA protein allows it to bind and carry four oxygen molecules. When the oxygen is released, that hemoglobin in the red blood cell retains its structure and is able to move along and pick up more oxygen. In sickle cell HbS, the one amino acid difference gives the protein a different shape from HbA. It can still bind to four oxygen molecules, but when it unloads the oxygen, the HbS molecules bind to each other and produce fibers which distorts the shape of the red blood cells, potentially resulting in a clot. The structure of all proteins determines their ability to function.

4.9 Questions for Review

4.9.1 *(Drawing of human hand in a couple formations---fist, thumb up, peace sign----probably this is in the text?)* Explain how this hand is a good analogy for the importance of protein structure.

In every chapter in your text, the following five sections appear as special boxes within the chapter. You should refer to those as you try to answer the review questions here, which focus on the main point(s) of the section.

4. Biology in Perspective

Sickle cell disease illustrates a number of important facts about genetics. One, a change in a single nucleotide in DNA can result in a changed amino acid that can result in a malfunctioning protein. In the case of sickle cell, normal hemoglobin, HbA becomes abnormal sickle cell hemoglobin, HbS, with serious health consequences. But it's not all bad since an individual who is merely a carrier of the HbS allele does not get full blown sickle cell disease and is resistant to an equally nasty disease, malaria. It's important to remember that genes do not act in isolation. They can affect each other and they can affect multiple characteristics. And the environment may exert considerable control over how genes behave. Your DNA alone does not make you who you are.

Question to consider:

Sickle cell disease can be deadly. Shouldn't a gene that causes a deadly disease disappear from the population because it kills nearly everybody that inherits it? Why hasn't this happened with the sickle cell gene?

4. Scientist Spotlight: Rosalind Franklin

James Watson, Francis Crick, and Maurice Wilkins shared the 1962 Nobel Prize for Physiology of Medicine for determining the structure of DNA. They could not have done this, though, without the input from another scientist, Rosalind Franklin. Through her work in X-ray crystallography she was eventually able to get a very good picture of the double strand of DNA. Unbeknownst to her, Wilkins showed her picture to Watson and Crick and they combined it with what they knew from their own work and accurately described the structure of DNA. Franklin died of ovarian cancer in 1958 at the age of 38 without ever really knowing what happened, and few people have heard of her.

Question to consider:

The work of Rosalind Franklin was instrumental in the discovery of the structure of DNA and yet few people have heard of her. Why is this?

4. Technology Connection: Electrophoresis

DNA is increasingly used to determine identity of individuals, often in crime scenes. A sample of DNA from saliva on a cigarette butt, for example, is collected. Scientists use certain enzymes like scissors to cut the DNA into fragments and run an electrophoresis gel. This can then be compared to gels of other DNA samples from known individuals. In this way, forensic scientists can identify the smoker of the incriminating cigarette butt.

4. Life Application: The Effectiveness of Genetic Screening

Because it is now possible to screen individuals for many genetic diseases, why do we still have people being born with these diseases? For example, some 8% of African Americans are sickle cell carriers even though it can be identified in DNA. Tay Sachs disease is another genetic disease, similarly identifiable and almost no one is born with this anymore.

Question to consider:

Tay Sachs disease has a genetic origin much like sickle cell, and screening for both is quite effective. In other words, when scientists examine the DNA of parents, they can identify if they are carriers of the disease and thus determine the risk of their children having the diseases. As a result of this type of screening, Tay Sachs is now very rare in North America. Why hasn't this worked for sickle cell?

4. How Do We Know?: Pedigree Analysis

A genetic counselor can ask questions about a disease that appears to run in a family. By mapping the family tree and indicating who has had the disease and who has not, using the rules of inheritance, a counselor can determine the history and at least some of the likely future of a genetic disease in a family. Many genetic diseases are also influenced by the environment, so heredity alone doesn't determine if an individual will get a disease, but can indicate risk.

Question to consider:

Pedigree analysis in looking at an individuals risk for a genetic disease is useful, but it isn't a sure thing. Why?

Student Study Guide

Chapter 5: Cancer
How can it be prevented, diagnosed, and treated?

Cancer is a disease of uncontrolled cell replication. It impacts the lives of most Americans either because they have the disease or they know someone who does. Close to 40% of all Americans will be diagnosed with cancer, and half of those will die from the disease.

Cancers have been described since the ancient Egyptian, Greek, and Roman empires. Some cancers are well understood, and some are known to be caused or exacerbated by certain behaviors. Since 1950, we've known that lung cancer is often related to smoking. Some cancers are genetically based.

In this chapter we look at what cancer is exactly, how it can be prevented, how it is diagnosed, and what treatment options are available.

5. Case Study: *Xeroderma pigmentosum*

5. Case Study Summary

Thirteen-year-old Jeff Markway finds a golf ball-sized lump on his arm and his doctor suspects it is a cyst. But in testing it, he determines that it is a melanoma, or skin cancer. It is unusual that a person so young would have this. Jeff is referred to Dr. Henry Lynch, a specialist in the study of genetic factors important in cancer. He determines that Jeff has *Xeroderma pigmentosum* (XP), a rare disorder in which people are extremely sensitive to ultraviolet (UV) light. XP increases the sensitivity of the skin to sunlight and speeds up the process by which UV light exposure causes skin cancer. Of the seven siblings in Jeff's family, five are found to have XP. Since XP is rare in the general population, there is clearly a genetic component to the disease.

Dr. James Cleaver, a scientist interested in the effects of UV radiation on DNA, offers to study the cells of this family to determine what goes wrong in XP. He finds that while normal cells exposed to UV radiation experience DNA damage, those cells are able to repair it to some extent. In XP cells, the damage is permanent, so it accumulates over a short time, making those cells more likely to become cancerous.

Questions to consider:

What does UV light do to cells?

What is the problem with XP cells (what can't they do?)?

What is the consequence of this for a person with the disease XP?

Learning Objective 5.1: How Does Cancer Make You Sick

> *Use appropriate terminology to describe cancer and the reasons why cancer makes you ill.*

5.1   Summary

You recall that a body is made up of cells which assemble into tissues. Tissues compose organs, which work together in organ systems. Each cell type has its own place and function in the body. They communicate, live, and die as they wear out or are damaged, then they are replaced. This process keeps a body running smoothly most of the time. Sometimes, however, cell birth and death becomes unbalanced and extra cells are produced that might form into a lump or **tumor**. A growth or tumor like this can be **benign**, or harmless. A wart, in which the cells don't keep growing and don't spread around the body, is an example of this. But a growth can also be **malignant**, meaning cell division is uncontrolled, and this is cancer.

Production of extra cells causes serious health consequences, even death, for three reasons:

1. *Tumors can grow quite large.* They impair organ function and may press down on blood vessels, impeding blood flow. Organs deprived of oxygen and nutrients will die.

2. Cancer can become invasive. Cancer cells may leave a tumor, enter the bloodstream, and spread new tumors anywhere in the body. This is called metastasis.

3. Tumors can deprive other cells of oxygen. Tumors secrete chemical signals that cause blood vessels from the circulatory system to grow into the tumor, thereby providing it with nutrients. This is called **angiogenesis** and because the tumor is receiving more oxygen and nutrients from the body, the normal healthy cells in the body receive less and can't function.

All cancers are simply this uncontrolled proliferation of cells, but there are more than 100 different types. They all involve some breakdown in the normal pathways that control cells and regulate the body.

5.1 Define these Key Terms

tumor

benign

malignant

metastasis

5.1 Questions to Review

5.1.1 What is cancer?

5.1.2 What is the difference between a benign and a malignant tumor?

5.1.3 Complete the following table:

Consequences of cancerous cells	What is this?	Why is this a problem for the body?
Tumors		
Metastasis		
Angiogenesis		

Learning Objective 5.2: How Do Cancer Cells Differ from Normal Cells?

Contrast the behavior of normal and cancerous cells.

5.2 Summary

Normal cells communicate with each other through chemical signals that influence many cell behaviors: cell division, cell migration, cell differentiation, programmed cell death. When one runs into another, they both stop moving and dividing so they don't pile up. Normal cells also have specific shapes or morphologies, they connect to each other in an orderly way to form tissues, and they have finite lifespans.

Cancerous cells behave as if they've lost the ability to understand chemical signals from other cells. They don't stop dividing, and they don't die. They lose the characteristics unique to the tissue from which they originated, and they often lose their physical connection to that tissue. They may detach and secrete enzymes that break down other cells. These enzymes may burn holes in the tissue, allowing the cancer cells to get into the blood stream and around the body, forming secondary tumors in new locations.

In a lab situation, in a culture, cancer cells never stop dividing, so they pile up. Sometimes the cells are actually immortal. They will keep dividing indefinitely.

5.2 Define this Key Term

morphology

5.2 Question for Review

5.2.1 Make a list of the behaviors and characteristics of a normal cell, and contrast that to the behaviors and characteristics of a cancerous cell.

Learning Objective 5.3: What is the Life Cycle of a Cell?

Use a signal transduction pathway to explain how cells become cancerous.

5.3 Summary

Cancer cells demonstrate a breakdown in the normal cell communication mechanisms controlling cell division, so the cells divide out of control. This can go wrong in three ways:

1. Normal cells require signal molecules called growth factors in order to grow. Growth inhibitors are signal molecules that prevent cell division. Cancer cells grow in the absence of growth factors, make their own, or are insensitive to growth inhibitors.

2. The signal molecule that is a growth factor triggers cell division by binding to a receptor in the cell membrane. This activates the signal transducer in the cell which relays the signal, molecule to molecule, until it reaches the transcription factors. These regulate gene activity and therefore control the genes regulating cell division. (Remember from Chapter 4 that transcription is the process of producing RNA in order to make proteins the cell needs.)   In cancer cells, the receptors, transducers, and/or transcription factors can be permanently activated. How this happens depends on the type of cancer, but basically the cells bypass the normal regulatory pathway and divide out of control.

3. Normal cells become damaged from age, wear and tear, or accidents. Sometimes enzymes can repair the damage, but if not, normal cells undergo apoptosis, or programmed cell death. Cancer cells sometimes lack the ability to make repairs (that's the case in Xeroderma pigmentosum), or they may ignore the signal to make repairs, so they keep dividing, accumulating errors, and grow increasingly abnormal.

The Cell Cycle

Cells ordinarily divide regularly in order to replace worn out cells, produce sperm, heal wounds, and grow babies. Because this is an important process, it's well regulated. Normally, the cell cycle, or cell division, consists of four distinct phases (refer to Figure 5.9):

1. Mitosis (M): The chromosomes separate into two daughter cells and the cell divides. (Ch. 3)

2. Gap 1 (G1): The cell grows and prepares for the next phase.

3. DNA synthesis (S): All the chromosomes replicate. (Ch. 4)

4. Gap 2 (G2): The cell grows and prepares again for mitosis.

Steps 2, 3, and 4, G1, S, and G2, constitute interphase which is what happens between divisions.

Under normal conditions, a series of "checkpoints" regulate the timing and amount of cell proliferation, for example, only allowing it to proceed if there is enough raw material to build a new cell, or the chromosomes have divided properly. This way only quality cells are produced as needed. Many cancer cells ignore the checkpoint controls and proceed through the cell cycle regardless of conditions.

5.3 Define these Key Terms

signal molecule

growth factor

growth inhibitor

signal transducer

transcription factor

apoptosis

mitosis

5.3 Questions for Review

5.3.1 There are three different ways cancer cells may come to divide out of control. In the table, describe what happens in each case.

Normal Controller	How a normal cell uses this	How a cancer cell works differently
Growth factor/inhibitor		
Signal transducer		
Apoptosis		

5.3.2. Why do normal cells need to divide?

5.3.3. What are the four steps in a normal cell cycle?

5.3.4. What goes wrong in the cell cycle of a cancer cell?

Learning Objective 5.4: In What Ways is Cancer a Genetic Disorder?

Describe the role that mutations play in cancer development.

5.4 Summary

All cancer cells have altered or mutated DNA compared to normal cells, so cancer is a genetic disorder in that sense. The genes in the cells don't function normally. A protein might be overproduced, abnormal, or not produced at all. If it is a protein important in cell communication or regulating division, the mutant cell may be transformed into a cancer cell dividing out of control. The mutations get passed to daughter cells and damage accumulates. Mutations of certain types of genes make a cell particularly vulnerable to becoming cancerous.

Oncogenes and Tumor Suppressor Genes

Whether or not a cell divides is normally precisely controlled by a network of proteins. The genes that produce these proteins are proto-oncogenes, and they work like the accelerator in a car, urging cells to divide. The brakes are tumor suppressor genes which tell the cell to stop dividing. Normally, the proto-oncogenes and tumor suppressor genes balance each other out. Proto-oncogenes can mutate to form oncogenes, which are like an accelerator stuck "on." Similarly, tumor suppressor genes can mutate so they no longer function and the "brakes" on cell division don't work. For a cell to be cancerous, at least one tumor suppressor must be disabled and several oncogenes activated.

For example, in certain types of leukemia or blood cell cancer, the tumor suppressor gene *NF-1* is mutated such that it can't do its job of inhibiting the proto-oncogene *ras*. With no brakes on *ras*, uncontrolled cell proliferation occurs.

At least 100 oncogenes have been discovered so far, as well as a few dozen tumor suppressor genes. Knowing the oncogenes and tumor suppressor genes involved in particular cancers provides valuable information for diagnosis, predicting outcomes, and treatment. For example, in about 30% of breast cancers, the oncogene *erB-2* overproduces a growth factor receptor, erB2-protein, on the breast cell membrane. The drug Herceptin binds to the erB2-protein and stops it from triggering cell division, thus controlling the cancer growth.

Chromosomal Abnormalities

When chromosomes, or parts of chromosomes, break, rearrange, stick together, or get lost, the result is major alterations in inherited instructions in DNA. These are large-scale mutations and if they turn proto-oncogenes into oncogenes or disable tumor suppressors, cancer can result.

The first case where a specific chromosomal alteration was found responsible for a particular cancer was the *Philadelphia Chromosome*. In 1960, Drs. Nowell and Hungerford found that patients with *chronic myelogenous leukemia* (CML) had an unusual, small chromosome in their cells, and normal healthy people did not. In the 1970s, Dr. Rowley was able to look more closely at the Philadelphia

Chromosome and found that chromosomes 9 and 22 had exchanged pieces, resulting in the Philadelphia chromosome. In 1984, Dr. Grosveld used molecular biological techniques to show that the chromosome rearrangement had repositioned two proto-oncogenes next to each other such that they were permanently stuck "on," resulting in uncontrolled cell division. So far at least 200 specific chromosome abnormalities like this have been consistently linked to particular cancers.

5.4 Define these Key Terms

proto-oncogenes

tumor suppressor genes

oncogenes

5.4 Questions for Review

5.4.1 Explain how normal proto-oncogenes and tumor suppressor genes act together to control cell division.

5.4.2 What is an oncogene and how does it produce cancerous cells?

5.4.3 Explain how the drug herceptin works to control breast cancer.

5.4.4. In general, how can a chromosomal abnormality result in cancer? The Philadelphia Chromosome offer an example of this which might help you explain it.

Learning Objective 5.5: What Risk Factors are Associated with Cancer?

Identify actions you can take to reduce your risk of cancer.

5.5 Summary

At least two thirds of cancers are influenced by factors we can control. These factors include smoking, poor diet, excessive alcohol use, lack of exercise, and excessive exposure to UV radiation. The factors we can't control include infections, carcinogens encountered in the workplace or in environmental pollution, genes, age, and wealth. We'll look at a few of these in more detail.

1. Smoking

For hundreds of years, it was suspected that there was a link between tobacco use and cancer, but it wasn't until 1948 when a medical student, after seeing a lifelong smoker's lungs at an autopsy, studied it carefully. He found that cigarette smokers were 40 times more likely to develop lung cancer than non-smokers. Smoking is responsible for at least 30% of all cancer deaths, and not just lung cancer, although 80–90% of lung cancers can be attributed to smoking. Tobacco is responsible for as many as 4 million deaths worldwide, of which about 40% are due to lung cancer. Public health education is reducing cigarette smoking in the United States, but it is still very common in China, India, Eastern Europe, and Africa.

2. Diet and Exercise

Smoking accounts for at least 30% of all cancer deaths in the United States, and another 35% can be attributed to diet. The problems are eating too much red meat and saturated fat, too few fruits and vegetables, and too little fiber. This is easy to see by looking at other countries. Japanese women in Japan have less than half the incidence of breast cancer seen in American women. (The incidence of breast cancer in the United States is about 12% of women.) Within one generation of a Japanese woman moving to the United States, however, her risk becomes just as high. In countries where the diet is rich in fruits, vegetables, and fiber, the incidence of colon and rectal cancers are ten times lower than in countries where the consumption of meat and fat is significant. Switching to a healthy diet lowers your cancer risk dramatically.

For exercise, the most critical period is childhood. A lack of exercise and extra body fat in childhood are responsible for 5% of prostate and breast cancers in adults.

3. Excessive Alcohol Use

Moderate alcohol use does not seem to increase cancer risk, but there is a link between heavy alcohol use (four or more drinks per day for an adult male) and cancers of the mouth, throat, liver, larynx, and esophagus, and possible breast cancer. Combining smoking and drinking magnifies the effects of both.

4. Radiation

We are routinely exposed to radiation from the sun, medical procedures, communication devices, and even our own bodies. Excessive exposure to high-energy radiation like x-rays can cause cancers, especially leukemia, but the risk is quite low. Electromagnetic radiation from radios, phones, cell phones, wireless systems, television, and electricity do not appear to increase cancer risk to a measureable degree. The real problem is ultraviolet radiation.

UV rays, found in sunlight and tanning beds, account for more than 90% of skin cancers. The most deadly form of skin cancer, melanoma, has increased dramatically since 1930 when data collection began. Melanoma is the leading type of cancer in young adults. It is an aggressive, metastatic cancer, diagnosed in 50,000 people each year, and it kills about 7000 of those.

5. Infection

Viruses are the most common type of infectious agent responsible for cancer, but at least one type of bacteria is associated with stomach cancer, and there are flatworms responsible for bladder and liver cancers. Whether these are a problem depends on the relative degree of affluence and poverty, as these can generally be prevented with medical screening.

The human papilloma virus (HPV) offers a good example of this. It is a sexually transmitted virus responsible for most of the world's cancers of the genitals and anus. Risk is directly related to the age at which a person becomes sexually active, and the number of sexual partners. The prevalence of HPV is higher in developing nations than in developed ones. It is responsible for 80–90% of cervical cancers worldwide. Cervical cancer is the second most common cancer affecting women, worldwide, but it is rare in the United States. This is due to the routine use of the Pap test, in which cervical cells are collected and screened for early signs of cancer. Also, in 2006, a vaccine which prevents infection from certain types of HPV was approved for use.

6. Workplace Carcinogens and Pollution

The World Health Organization's International Agency for Research in Cancer has identified more than 40 specific materials in the workplace which are associated with increases in particular cancers. Exposure to these materials at work is responsible for 5% of all cancer deaths in the United States. Implementation of safety measures has reduced this from 10% in 1950 in the United States, but rapid industrialization in developing nations will likely keep this incidence relatively high.

Scientists estimate that 2% of all cancers can be attributed to air, water, and/or soil pollution.

7. Inheritance

Most cancers display no family history at all, but in about 5–10%, an inherited mutation may be involved. You saw an example in Ch. 4 in which breast cancer was prevalent in a family pedigree, suggesting a genetic factor. Scientists have indeed found two gene mutations, *BRCA-1* and *BRCA-2*, which account for 90% of all hereditary breast cancers. A woman with these genes faces an 84% probability that she will develop breast cancer by age 70.

8. Age

Your likelihood of contracting cancer increases as you age. Remember that cancer is the result of defective DNA in cells, and the more cells divide, the more they will accumulate errors and damage. The longer you live, the more likely it becomes that your cells will accumulate enough damage to become cancerous.

9. Poverty

The link between poverty and cancer is found in all countries. There are several reasons. One is that the risk factors for cancer tend to be more prevalent among the poor: smoking, poor nutrition, exposure to carcinogens in the workplace, and exposure to infectious agents (sometimes from lack of clean water). Second, poor people lack access to health care, so they are not screened for cancer, and later diagnoses are more difficult and expensive to treat.

5.5 Define these Key Terms

melanoma

HPV

Pap test

5.5 Questions for Review

5.5.1   Make a list of things you could do to lower your risk of getting cancer.

5.5.2 Make a list of those cancer-causing factors you can't control.

5.5.3 If cancer repeatedly appeared in your family history, what might you conclude about the cause of that particular cancer?

5.5.6 What does Figure 5.19 tell you? Why is this the case?

5.5.7 Why are poor people at higher risk of getting cancer?

Learning Objective 5.6: How is Cancer Diagnosed?

Explain how cancers are diagnosed.

5.6 Summary

You know your body best, so you are a doctor's most important source of information when it comes to diagnosing cancer. Usually, something that doesn't seem right will not be as serious as cancer, but regular checkups are important. Table 5.2 lists the American Cancer Society guidelines for early detection for the most common cancers. You should visit a doctor if you notice any of the following:

- A lump in your breast, testes, or anywhere else
- Changes in the color or size of a wart or mole
- Changes in bowel or bladder habits
- Persistent indigestion or heartburn, or difficulty swallowing
- Unusual bleeding

Mammograms, a form of breast x-ray, are advisable annually, for women 40 and older. Pap tests to screen women for cervical cancer should begin by age 20. Beginning around age 50, men and women should get a colonoscopy to screen the entire rectum and anus for potentially cancerous growths. A doctor will also examine the testes and prostate gland physically to check for lumps. Anything suspicious in any of these screenings warrants further testing.

Further tests may include blood tests to look at the blood cells or measure levels of particular proteins. The PSA test, for example tests for high levels of prostate-specific antigen, an indication of prostate cancer. Ultrasound, computerized tomography (CT scans), or magnetic resonance imagery (MRI) provide the doctor with a detailed look at your internal organs. Finally, a sample of the suspect cells, or biopsy, will be taken to see if they are cancerous.

Of course, doctors don't want to miss anything, so these tests are as broad as possible. Sometimes there are "false positives." About 10% of women who get mammograms get called back for follow-up tests that turn out to be negative. This causes unnecessary
worry, but it's better than "false negatives," in which doctors miss the diagnosis.

5.6 Define these Key Terms

mammogram

colonoscopy

PSA test

false positive

5.6 Questions for Review

5.6.1 List a few body changes for which you should see a doctor.

5.6.2 Why do you have to take the lead in cancer diagnosis, rather than just leaving it up to your doctor?

5.6.3 How are people screened for breast, cervical, and colorectal cancers?

5.6.4. If something suspicious shows up in a screening, does that mean you have cancer? What happens next?

5.6.5 What is a false positive in a cancer screening?

Learning Objective 5.7: How is Cancer Treated?

Contrast the three methods used to treat cancer.

5.7 Summary

There are three principal methods used to treat cancer, often used in combination.

1. Surgery

Surgery is performed on more than 50% of cancer patients, to try to remove the cancerous tissue, or in performing the biopsy. This is most successful when the tumor is localized and has not metastasized (spread). Pathologists, doctors who specialize in the study of disease, will then examine the tumor to determine just how abnormal the chromosomes and cells are. This provides guidance for other possible treatments.

2. Radiation therapy

Radiation therapy is used to kill cancer cells that could not be removed by surgery, or to shrink a tumor prior to surgery. This can be done in two ways:

> a. External radiation therapy: The tumor is bombarded with a focused beam of x-rays. (Fig 5.23)

> b. Internal radiation therapy: A localized source of radiation or "seed" of radioactive material is placed near or in the tumor in order to try and kill the cells.

3. Chemotherapy

Chemotherapy is used to treat cancers that have spread from the primary tumor such that surgery would be impossible. Sometimes it's used if there is concern the tumor has metastasized, even if that is not clearly observable. It consists of drugs taken orally or injected directly into the blood stream. The drugs impair the ability of the cells to replicate, ultimately killing them.

Unfortunately, sometimes even the most aggressive cancer treatments fail. Cancer cells change, and each new round of cell division can produce increasingly abnormal or invasive cells. Natural selection will select for the cancer cells most resistant to treatment, to the point where a drug is ineffective. New drugs can then be tried, but ultimately, nothing may work and the patient will die.

5.7 Define these Key Terms

pathologist

radiation therapy

chemotherapy

5.7 Questions for Review

5.7.1 Fill in the table with information about the methods of treating cancer.

Method of Treatment	When is this used?	What is the goal?
surgery		
Radiation therapy		
Chemotherapy		

5.7.2. Why does cancer treatment sometimes fail?

Learning Objective 5.8: How Can Cancer Be Prevented?

Enumerate steps to decrease cancer risk.

5.8 Summary

Many cancers are preventable for most people. At least 65% of all cancers in the United States are due to smoking, poor diet, and lack of exercise. Excessive exposure to sunlight is the main cause of skin cancer. Promiscuous sexual activity puts you at risk for a cancer-causing pathogen. Clearly, the decisions you make about your lifestyle can modify your risk for cancer.

As an example, consider smoking. A British study verified the link between smoking and lung cancer in the 1950s. In 1955, the lung cancer incidence of British men was the highest in the world. Once the public finally caught on to this, things changed. British cigarettes were designed to have lower tar levels, as cigarette tar is a potent carcinogen. A public health campaign succeeded in lowering the number of people who smoked. The lung cancer incidence in British men dropped below 65%. Prevention works!

In fact, scientists estimate that at least ten times more lives could be saved if we focused on prevention rather than treatment. Scientists also estimate, though, that 20–25% of the cancers that occur now would continue even if we eliminated all external causes of cancer. Some are inevitable, and some happen by chance.

5.8 Questions for Review

5.8.1 Fill in the table of simple steps you can take to reduce your chances of cancer. (refer to Table 5.3 in the text)

5.8.2 Why might cancer prevention be just as important as cancer treatment?

5. Biology in Perspective

Cancer is a disease of uncontrolled cell proliferation, a consequence of mutations of the genes responsible for regulating cell division. Many factors increasing a person's risk of developing cancer are avoidable: smoking, poor diet, exposure to infectious agents, exposure to radiation (UV light). Some factors are difficult or impossible to avoid: pollution, workplace carcinogens, age, family history, poverty.

Cancer survival is dramatically helped by early diagnosis through physical exams, blood tests, and visualization tools like ultrasound, CT scanning, and MRI. Treatments include surgery, radiation therapy, and chemotherapy. Access to medical care makes early detection more likely, of course, and this is one reason poverty increases cancer risk.

Question to consider:

Scientists know a lot about cancer prevention and treatment, yet many people still get cancer and die of the disease. What are a few of the reasons for this?

5. Scientist Spotlight: Peyton Rous (1879–1970)

Peyton Rous got a medical degree from Johns Hopkins University and went into medical research at the Rockefeller Institute in 1909. That year, a farmer from Long Island brought Dr. Rous a prized hen, hoping he could treat a tumor in her right breast muscle. Rous ended up doing two experiments with the bird. First he removed the tumor and transplanted it to another chicken. That chicken also developed cancer, demonstrating that cancer could be transferred. Then Rous ground up the tumor tissue and collected the liquid that squeezed out. He filtered the extract to remove all debris and cells, and injected it into another chicken. Again, the chicken developed cancer, demonstrating that the infectious agent was smaller than a cell, and too small to be seen with the microscopes of 1909. We now know it was a virus, now named Rous sarcoma virus.

Unable to produce tumors in mammals by exposing them to the virus, Rous thought it was unique to chickens, and therefore not generally significant. Of course the discovery of cancer-causing viruses has been very significant, and Rous was awarded a Nobel Prize for his work in 1966 when he was 87 years old.

Question to consider:

Why didn't Dr. Rous initially think his work with a cancer virus was of any use? (Hint: Why did Dr. Rous try to infect mammals with the virus?)

5. Technology Connection: Computerized Tomography (CT Scans)

Standard X-rays penetrate skin and soft tissue, but some parts of the body are more opaque and absorb X-rays, showing up on film as dark areas. Bones show up this way, and often tumors are also denser than surrounding tissues and show up darker. Computerized tomography, or CT scans are a more advanced type of X-ray that provides a more detailed look at internal structures, including tumors. They send the X-ray beams into the tissue at different angles, rather than a straight beam, which results in a three-dimensional image. A doctor or surgeon gets a much clearer, three-dimensional picture of the tumor this way.

Question to consider:

What is the difference between a tumor as you see it with standard X-rays *versus* with a CT scan?

5. Life Application: Chemoprevention

Many diets have been marketed that claim to provide health benefits, but do they? We know some foods increase our chances of developing cancer. Are there things we could eat that would reduce our risk? Beginning in the 1950s, scientists looked at whether natural or manufactured materials could prevent or halt cancer development. This is chemoprevention.

Research has shown a diet rich in fruits, vegetables, and fiber, and low in saturated fats decreases the risk of developing certain cancers, but are the phytochemicals, components derived from plants, are responsible for the effect? A phytochemical called dithiolethione, found in cruciferous vegetables such as broccoli, cabbage, and cauliflower, has been shown to inhibit the development of lung, colon, mammary gland, and bladder tumors in lab animals. Studies suggest that dithiolethione interferes with cancer formation by activating liver enzymes that detoxify carcinogens. There are many more materials being investigated which may prove to be helpful.

Question to consider:

What is chemoprevention and what types of materials are being researched as possibly helpful in preventing cancer?

5. Define these Key Terms

chemoprevention

phytochemicals

5. How Do We Know?: Cancer-Causing Genes from Malfunctioning, Normal Genes

By the late 1970s, scientists knew that certain viruses carried specific genes which could transform normal cells into a cancerous state, but they also knew that these same cancer genes could be found in normal cells too. Was cancer due to stray genes introduced by a virus, or could previously normal genes simply malfunction?

To address this, Dr. Chaio Shih and Dr. Robert Wienberg conducted an experiment. They cultured normal cells in the lab and treated them with a known carcinogen, or cancer-causing substance. The cells became cancerous. Next they isolated the DNA from the cancerous cells and inserted it into a second set of normal cells. Those cells, too, became cancerous, demonstrating that it was the DNA that transformed them; no virus necessary. Would tumor cells from a person, rather than from a lab culture, also transform normal cells to cancerous cells? To test this, Dr. Weinberg and Dr. Geoffrey Cooper isolated DNA from a human bladder cancer and inserted it into normal cells in culture. The normal cells again became cancerous, demonstrating that cellular DNA can be altered such that cancer will result.

Question to consider:

How did scientists show that cancer could be caused simply by introducing altered cell DNA to a healthy cell?

Student Study Guide

Chapter 6: Reproduction

What Kind of Baby Is It?

Producing offspring is the most important thing we do as a species, so understanding the biology of reproduction is essential for understanding what makes us human. It is also essential for both responsible family planning and stewardship of our natural resources. It is all about communication between multiple entities: individual cells, eggs and sperm, embryo and uterus, and fetus and mom. So does the ability to give birth and the ability to father a child define a woman and a man? It might not be that simple.

6. Case Study: The Fastest Woman on Earth

Caster Şemenya came from Limpopo, a province in South Africa. In 2008, she won a gold medal at the Commonwealth Youth Games and then the African Junior Athletics Championship and qualified for the 2009 World Championship at age 18. But then, because she looked masculine and ran so much faster than her competitors, some people contended that she must be male. She was subjected to gender tests.

Semenya passed the tests, but the harassment continued as she crushed opponents in the 800 meters to win the world championship. It turns out the International Association of Athletics Foundation (IAAF) had leaked the rumor that Semenya was not a woman. Australia's Daily Telegraph claimed test results showed that she had external female genitals but no ovaries or uterus, undescended testes, and testosterone levels 3X higher than the average woman, although still much lower than a man's. The IAAF allowed her to keep her gold medal and she waited for a year to compete again, waiting for the IAAF decision on her eligibility. In 2010 she was declared free to compete professionally with women. She won a silver medal at both the 2011 world championships and the 2012 Summer Olympics in London, and carried the South African flag in the opening ceremonies at those Olympics.

But how should the IAAF answer this difficult, if not impossible question: What is the ultimate difference between males and females? Is there something that can be measured or tested that shows unequivocally that someone is a man or woman? Can you be neither?

Question for Review

Who is Caster Semenya and even though she was a girl as far as she knew, why did some people suspect she might not be?

6.1 Learning Objective: How do males and females function?

Describe how chromosomes and hormone instructions interact to determine sex.

6.1   Summary

As many as 1.7% of the population has a disorder of sexual development or DSD, which means that their development is profoundly different from the standard definition of male and female. Yet it's possible to go through life with a DSD and not even know it. Why is there so much variability?

The Stages of Sex Determination

Sex determination begins with the gametes, the egg and the sperm that unite to form a new organism. From there, in some lizards and turtles, the sex of the offspring is determined by the temperature at which the eggs incubate. In the marine worm *Bonellia*, the sex of the newly hatched young depends where they settle on a solid surface. If that surface is the front of a female worm, the young worm will develop as a male. It it's some other surface, it will develop as a female.

In some cases, sex isn't permanent. A male clownfish, *Amphiprion ocellaris*, changes to a female later in life. A black sea bass *Centropristis striata* starts life as a male and can become female later.

Primary Sex Determination

Humans and other mammals have two types of sex chromosomes: X and Y. Normal cells have two sets of chromosomes so males are XY and females are XX. When gametes are formed for reproduction, males make sperm cells with either an X or a Y chromosome and females make eggs with one X chromosome each. When egg and sperm fuse, the resulting zygote has an X from the mother and either an X or Y from the father.

An embryo with XX will develop into a girl with ovaries. An XY means the embryo will develop into a boy with testes. This is primary sex determination.

Secondary Sex Determination

This governs the development of secondary sexual characteristics typical of males and females. Males have external genitals. Women have an external vulva and breasts, and an internal uterus and vagina. Men and women also differ in pelvic structure, voice tone, and locations of body fat and hair. Chemical signals called hormones are responsible for this development.

Chromosome Instructions

Before an embryo is 6 weeks old, it is not physically male or female. The structures in the embryo that will become internal sex characteristics are all there, but they await instructions from genes. A normal Y chromosome has the SRY gene that is necessary for the development of testes. If a mutation disables the gene, the embryo will develop ovaries, even if it's XY. If an XX embryo has the SRY gene because of a genetic error, testes develop. The X chromosome has a gene called DAX1, and two of these genes are required for normal ovary development. If the female has just one X, she won't develop normal ovaries.

Hormone Instructions

The early embryo has "plumbing" in place for eventual development of either male or female internal sex characteristics. Once the ovaries or testes form, they secrete hormones regulating which structures stay and which disappear. For example, XY and XX embryos have both Wolffian ducts and Mullerian ducts. In a XY embryo, anti-Mullerian hormone triggers the demise of the Mullerian ducts and testosterone promotes development of the Wolffian ducts, the prostate gland, and the penis. It also inhibits breast formation and regulates descent of the testes into the scrotum. In the XX embryo, the ovaries secrete estrogen which promotes the development of the Mullerian ducts, oviducts, uterus, and upper end of the vagina. Ovaries also secrete some testosterone, just not as much as testes do.

At 6 weeks, the embryo also has indistinguishable external sex characteristics, with just a small bud between its legs. That will be the penis or clitoris. By 9 weeks, there are swellings on either side of the urogenital groove. By 14 weeks, the groove disappears in male embryos and the swellings become the scrotum. In females the groove becomes the vagina and the swellings the external folds of the female genitals, the *labia majora* and *labia minora.* All these changes are regulated by hormones, which also increase in puberty to enhance the differences between males and females.

6.1 Define these Key Terms

disorder of sexual development

ovaries

testes

primary sex determination

secondary sex determination

hormones

testosterone

estrogen

sex chromosomes

6.1 Questions for Review

6.1.1 Draw a flow chart illustrating the development of sperm and egg cells from the parents to when they join in the offspring. Indicate the sex chromosomes.

6.1.2 Are sex chromosomes the only way gender is determined in all developing animals? Explain with examples.

6.1.3 If an embryo gets XY chromosomes, how can a genetic mutation still alter the development of sexual characteristics. What if an embryo just gets one X chromosome? What about an embryo with XX but also the SY1 gene?

6.1.4 Draw a flow chart illustrating the development of sex characteristics in an XX and an XY embryo, from fertilization forward. Indicate which hormones regulate what.

Learning Objective 6.2: What happens if the hormonal signals are missing or misread?

Predict the outcome of a change in hormone production or response.

6.2 Summary

About 2% of men and women have one of more than 40 different types of disorders of sexual development. Your text discusses two of these.

1. Androgen-Insensitivity Syndrome

In Androgen-Insensitivity Syndrome (AIS), individuals who are XY do not develop as males because their cells lack receptors to bind to testosterone. The SRY gene functions correctly and testes develop. The testes produce testosterone so the Mullerian ducts deteriorate, but no external male genitals form so the testes remain internal.

The Mullerian ducts are destroyed, so no internal female body parts form either, but the AIS embryos are exposed to estrogen from the mother. This results in breast development and the female pattern of hair and fat. So the AIS person will look and be female, even though she is XY. She may be completely unaware of the problem until she tries to find out why she isn't menstruating or is unable to become pregnant.

AIS can sometimes be partial, when just a bit of testosterone gets through. The woman may be XY and appear female but may also have some male characteristics like greater muscle mass, little breast development, and a less female-shaped pelvis. Possibly, Caster Semenya had partial AIS.

2. Pseudohermaphroditism

In pseudohermaphroditism, an XY individual develops testes but is missing an enzyme that lets testosterone send the right signals, so the external genitals look female. Since the Mullerian duct degenerates, however, there are no internal female ducts. (The name of this DSD comes from a "hermaphrodite," which means having both male and female parts, and "pseudo" which means apparently rather than actually.)

At puberty, the adrenal gland produces testosterone. What seemed like a clitoris enlarges into an obvious penis. The voice deepens, muscles enlarge, breasts do not develop, and menstruation does not begin. These people start life as females and then develop male characteristics.

We've called these individuals with DSD male or female, but maybe that designation should be left open for this group.

6.2 Define this Key Term

adrenal gland

6.2 Questions for Review

6.2.1 What goes wrong in AIS and what are the developmental consequences?

6.2.2 What goes wrong in psuedohermaphroditism and what are the developmental consequences?

Learning Objective 6.3: How do men produce sperm?

Outline the steps of sperm production in males.

6.3 Summary

It takes about 9–10 weeks to make sperm, and males do this continuously from puberty practically until death, although the quality does deteriorate with age.

The Testes

Sperm is produced in the testes, which are about 5 centimeters long and located in the pouch called the scrotum. Sperm is very sensitive to temperature, so the testes are held in place with small muscles which move them closer to the body in colder situations. Testes are very sensitive in general, so it's risky to have them hanging out and vulnerable, but sperm also don't work if they're too warm, and inside the body is too warm.

The testes consist of around 30 wedge-shaped units, each containing three tightly packed coils of 125-meter-long seminiferous tubules. (Remember that 1 meter is 39 inches.) Lining the walls of the tubules are self-renewing stem cells. When these undergo mitosis (see Ch. 3), one new cell replenishes the stem cell population and the other is a primary spermatocyte. The spermatocyte undergoes meiosis I, producing secondary spermatocytes. These undergo meiosis II to form the spermatids which differentiate into sperm.

Sperm and Semen

Sperm initially leaving the testes could not fertilize an egg. They move into the **epididymis** which secretes enzymes necessary to complete differentiation into working sperm. The epididymis is a 6-meter-long coiled tube, sitting atop each testes. Sperm are stored here until the male is sexually aroused, at which point they speed into the vas deferens, past ejaculatory ducts, and into the urethra which runs through the penis. In ejaculation, they're propelled out of the body.

Because sperm swim, they need liquid to move. The urethra also carries urine, so before ejaculation, two glands secrete a clear fluid into the urethra to neutralize any urine residue. The **prostate gland** secretes a milky fluid that also neutralizes the acidity of the vagina so the sperm survive. Finally, two seminal vesicles secrete the sugar fructose to provide energy for the sperm, and **prostaglandins**, a chemical signal that causes muscle contractions in the female reproductive tract that gets the sperm closer to the egg. The sperm plus these secretions is called **semen**.

Hormones and Sperm Production

The job of a sperm cell is to fertilize an egg to make a zygote. Not enough sperm could lead to low fertility. Too many may mean problems with quality. Several hormones control the process of making sperm.

First, the **hypothalamus**, a part of the brain, secretes gonadotropin releasing hormone (GnRH), which stimulates the release of the follicle stimulating hormone (FSH) and the luteinizing hormone (LH) from the **pituitary gland**, also in the brain. In males, FSH stimulates sperm production and LH triggers testosterone production in the testes. These hormone levels are monitored and balanced by the hypothalamus, which can also shut the whole process down if testosterone levels get too high by inhibiting GnRH release. If the sperm count gets too high, Sertoli cells release inhibin that both deters GnRH release and stops FSH release.

With all these hormones involved, molecules such as pollutants or drugs which disrupt normal hormone signals can have serious consequences for sexual development, secondary sex characteristics, and fertility. Anabolic steroids are known to do this, as is a pesticide called atrazine.

6.3 Define these Key Terms

stem cells

epididymis

prostate gland

prostaglandins

semen

hypothalamus

pituitary gland

6.3 Questions for Review

6.3.1 Why are the testes external to the body?

6.3.2 Draw a flow chart of sperm development from stem cells to sperm to ejaculation. Include input from any glands along the way.

6.3.3. Diagram the pathways by which the hypothalamus controls sperm production. Which hormones do what?

Learning Objective 6.4:   How do women produce eggs?

Outline the steps of egg production in females.

6.4 Summary

Unlike men who continuously produce sperm, women are born with all the immature eggs or primary oocytes they'll ever have, and from puberty until menopause, women only release one egg per month. But a woman's body also needs to prepare for pregnancy, possibly carry a pregnancy to term, and producing fertile eggs (oogenesis) is more complicated than producing sperm.

The Ovaries

Ovaries are the gamete-making organs in women. Tucked safely inside the body, these organs mature and release eggs and produce hormones important in the development of secondary sex characteristics. There are two ovaries. They alternate producing one egg per month and the egg is propelled into an oviduct, adjacent to each ovary. The egg is swept into the uterus by hair-like structures inside the oviduct. The uterus is about the size and shape of an upside down pear, has thick, muscular walls, and is where a fetus would develop. The cervix includes the lower part of the uterus and the opening to the vagina.

A female fetus develops about 2 million potential egg cells or oocytes. By puberty, a girl has about 700,000 of these left, and fewer than 500 are ever released for possible fertilization. Each oocyte starts meiosis I in the ovary in the fetus and pauses until puberty. Once puberty occurs, one oocyte completes meiosis I each month. One of the resulting cells, the secondary oocyte, starts meiosis II and about 12 hours later is released from the ovary: ovulation. Meanwhile the uterus is preparing for pregnancy each month. This entire process is controlled by hormones.

The Ovarian Cycle

At the start of the ovarian cycle, the oocyte completes meiosis I and begins meiosis II. Like with men and sperm, it's the hypothalamus secreting GnRH, triggering the pituitary gland to release FSH and LH, that controls oocyte development and release. Each primary oocyte is surrounded by a layer of cells called a follicle. FSH (follicle stimulating hormone) causes the follicle to grow, and with LH, stimulates estrogen secretion. Fluid builds up in the follicle until it bursts, releasing the secondary oocyte.

After ovulation, the body begins to prepare for pregnancy. The ruptured follicle develops into the corpus luteum that secretes estrogen and progesterone. Levels of these hormones in the blood get so high that no more follicles ripen for as long as the corpus luteum lasts. If there is no pregnancy after about 12 days, the corpus luteum breaks down and the cycle starts again.

The Uterine Cycle

Every month a woman is not pregnant, the uterus prepares its inner surface for pregnancy. This cycle consists of three phases:

> 1. Menstruation: This is the shedding of the uterine lining, the endometrium, which would have supported fetal development. Menstrual fluid is a combination of blood and endometrial tissue. Once it is gone, increasing estrogen levels stimulate development of a new one.

2. Ovulation: About half way through the 28-day cycle, estrogen prompts the cervix to secrete a thin, clear mucus easy for sperm to swim in. Ovulation occurs.

3. Luteal phase: The corpus luteum secretes estrogen and progesterone that maintain the endometrium. If there is no pregnancy, the corpus luteum degenerates, estrogen and progesterone levels plummet, and the endometrium breaks down.

Hormones and Pregnancy

For a developing egg, the journey from the ovary to the oviduct, uterus, and vagina takes about 3 days. Fertilization can happen anytime during this phase, but fertilization leading to a successful pregnancy almost always occurs in the oviduct. The embryo then travels down the oviduct and into the uterus and implants itself into the wall of the uterus. The endometrium does not break down and secretes human chorionic gonadotropin (hCG), which allows the corpus luteum to survive another 6–7 weeks, long enough for the placenta to develop. The placenta then makes enough progesterone to maintain the uterine lining for the duration of the pregnancy.

6.4 Define these Key Terms

oogenesis

primary oocytes

oviduct

estrogen

progesterone

corpus luteum

ovary

endometrium

ovulation

hCG

6.4 Questions for Review

6.4.1 Using a flow chart, diagram the path of an oocyte to mature egg, with hormone inputs noted.

6.4.2 Draw a graph with each stage of the ovarian cycle on the x-axis, and the levels of estrogen and progesterone on the y-axis. Describe what is happening in each stage.

6.4.3 Where does fertilization normally occur? Describe the next sequence of events that will maintain a pregnancy.

Learning Objective 6.5: How can pregnancy be prevented?

Compare methods used to prevent pregnancy.

6.5 Summary

The key to preventing pregnancy is to keep sperm and eggs apart. Birth control or contraception comes down to four methods:

1. Surgery

Surgery offers essentially a permanent end to fertility. For men, a vasectomy involves snipping and tying off the vas deferens, preventing sperm from entering semen. In a tubal ligation, the oviducts in a woman are cut and sealed so eggs cannot get into the uterus.

2. Hormones

A combination of estrogen and progesterone, delivered by implant, injection, vaginal ring, patch, or pill, can disrupt a woman's normal cycle. Pregnancy is impossible because either follicle development or endometrium formation is prevented.

3. Barrier

After abstinence, these are the oldest forms of birth control. The goal is to keep eggs and sperm apart, and there are a few varieties of barriers. In women, a cervical cap placed over the cervix or a diaphragm placed in the vagina near the cervix do this. Male condoms cover the penis to keep sperm from entering the vagina at all. Female condoms line the vagina completely, keeping sperm out of the uterus. Both male and female condoms also help prevent the spread of sexually transmitted diseases.

4. Other

Spermicides kill sperm cells and are most effective when used in combination with a barrier. An intrauterine device or IUD is inserted into the uterus by a physician and prevents the development of the endometrium and, therefore, implantation by an embryo. There is a vaccine against hCG in development, and future vaccines could target sperm itself.

6.5 Define these Key Terms

contraceptives

vasectomy

tubal ligation

condom

IUD

spermicide

6.5 Questions for Review

6.5.1 Fill in the following table with at least two methods in each category of contraceptive.

	Contraceptive Method	Male or Female?	How does it work?
Surgery			
Hormones			
Barrier			
Other			

Learning Objective 6.6: What causes infertility and how can it be treated?

Predict causes of infertility and propose potential treatments.

6.6 Summary

About 15% of all couples are infertile, which is defined as a failure to conceive and become pregnant after a year of regular, unprotected intercourse. About 40% of the time the problem lies with the man, 40% with the woman, and 20% a combination of both.

Causes of Fertility

1. Infertility in men: Low sperm count overall, or a low count of healthy sperm is usually the problem. Reasons for this are not clear, but it's correlated with smoking and alcohol and drug use. Injury to the testes may also make it difficult for the sperm to enter the vas deferens. Sperm counts are declining in men in developed, industrialized countries. It might be that a sedentary lifestyle or job that requires sitting elevates the temperature of the testes. It might be harmful pollutants.

2. Infertility in women: Sometimes the hormonal cycles don't work properly, so ovulation fails or the uterine lining is not prepared properly. In endometriosis, uterine tissue grows elsewhere in the body, usually the oviducts and ovaries. The tissue still responds to hormones, growing and being shed, but it is not easily eliminated from the body and can lead to painful inflammation and disrupted ovulation. Pelvic inflammatory disease (PID), caused by a sexually transmitted disease or an IUD-related infection, can also cause infertility.

In men and women, disorders of sexual development (as discussed in section 6.2) can cause infertility.

Infertility Treatments

Historically, the only option for an infertile couple to have a child was to adopt. In 1978, though, the first "test tube" baby was born, and that was the beginning of assisted reproductive technology (ART). The goal of ART is to bring healthy sperm and eggs together so fertilization is possible. There are five ART methods:

1. Artificial insemination: Sperm from a male donor is inserted into the female partner's vagina. If the man has a low sperm count, his sperm can be concentrated for this procedure, or sperm from another donor can be used.

2. Intrauterine insemination: Sperm is placed directly into the uterus. This is used if the man's sperm has motility problems (doesn't move well). It gives the sperm less distance to travel. Also sometimes the woman's vagina is too hostile an environment.

3. *In vitro* fertilization (IVF): This is the "test tube" baby. A doctor uses ultrasound to see developing follicles in an ovary. The immature oocytes are collected with a fine pipette and placed in Petri dish in a solution that permits them to mature. Sperm from the male partner or a donor are combined with the oocytes, and after 2–4 days, a small number of embryos are transferred to the uterus at the 8-cell stage in hope that at one will implant, resulting in a pregnancy.

4. Gamete intrafallopian transfer (GIFT): This is a variation of IVF in which the egg and sperm, or a zygote, is put into the oviduct.

5. Intracytoplasmic sperm injection (ICSI): A physician injects a single sperm cell into an egg, and if an embryo results, it is placed into the uterus no later than the 8-cell stage.

When none of these treatments work, a couple may opt for surrogacy, in which another woman gives birth to the baby after being artificially inseminated with the male partner's sperm, or having a "test tube" embryo implanted. Sometimes fertility drugs can work to stimulate the ovaries, or surgery may help in the case of a physical blockage to eggs or sperm.

The ART process can be difficult, painful, and expensive. It is not guaranteed to work, but more than 200,000 children are born each year as a result of ART treatments.

6.6 Define these Key Terms

infertility

endometriosis

assisted reproductive technology

6.6 Questions for Review

6.6.1 Low sperm count typically causes infertility in men. What may cause that?

6.6.2 What may cause infertility in women?

6.6.3 Fill in the following table:

Infertility treatment	What is the process?
Artificial insemination	
Intrauterine insemination	
In vitro fertilization (IVF)	
Gamete intrafallopian transfer (GIFT)	
Intracytoplasmic sperm injection (ICSI)	
Surrogacy	
Fertility drugs	

Learning Objective 6.7: How can we tell if a fetus or baby is healthy?

Explain the purpose and methods for testing fetal and newborn health.

6.7 Summary

Once a pregnancy is established, concern shifts to the health of the baby. There are many ways to assess this.

Blood and Urine Tests

With blood samples from the mother, the physician can determine blood type, and whether she is anemic or has been exposed to infections potentially dangerous to the fetus. Sexually transmitted diseases like HIV/AIDS, syphilis, and hepatitis are of particular concern. Rubella (German measles) or chickenpox also pose risk. Genetic tests may be done to look for potential genetic problems. High blood pressure can also endanger a pregnancy.

Nearly every trip to the doctor for a pregnant woman involves a urine test. The doctor monitors glucose levels to look for diabetes and protein levels which could indicate preeclamsia.

Screens and Diagnostic tests

Screening tests determine whether an embryo or fetus is at risk for a particular problem. A positive screen test may prompt a diagnostic test to determine whether there is a problem. Risk factors include genetic issues, problems with a previous pregnancy, or a mother's ill health.

Blood tests during the first and second trimesters can be used to assess fetal health because certain molecules in fetal blood, which would show up in the mother's blood, are known to vary in Down syndrome or spina bifida, among others. Ultrasound is used to measure the fetus and allows a doctor to detect physical problems.

If there is a concern, the next step is diagnostic tests which might include amniocentesis to examine fetal cells. This can be done between 14 and 17 weeks of pregnancy. A long needle is inserted into the uterus through the abdomen and fluid containing fetal cells is drawn out. These cells can be cultured and their chromosomes examined to make sure they look normal. They can also be tested for the presence of genes associated with hemophilia, muscular dystrophy, and cystic fibrosis.

In chorionic villus sampling (CVS), fetal cells are collected from the chorion, or fetal part of the placenta. This is somewhat riskier, but can be done between 10 and 12 weeks.

These tests increase the chance of a healthy birth but also raise difficult questions. Should you have the tests? What if a serious defect is detected? Possibly, treatments will lessen symptoms, and the parents can prepare for a child's special needs if they know in advance. Or they could choose to terminate the pregnancy. Suppose the fetus is found to have a disorder of sexual development?

Newborn Tests

Within 1–5 minutes of being born, a baby's health is assessed using what is known as the APGAR scale. This test, developed by Dr. Virginia Apgar records the baby's **a**ctivity, **p**ulse, **g**rimace **r**esponse, **a**ppearance, and **r**espiration. Each category is scored 0–2, and the baby is weighed and measured. A doctor or nurse can then tell immediately whether some help is needed. In the U.S., babies also get an injection of vitamin K, which helps in blood clotting, and eye drops to prevent infection. They may also get a hearing test.

Since the 1960s, blood samples have been taken; early detection can give physicians the chance to suppress or lessen symptoms of at least 30 diseases.

6.7 Define this Key Term

APGAR

6.7 Questions for Review

6.7.1 Explain the difference between a screening and a diagnostic test.

6.7.2 Fill in the following table:

Tests	Screening or diagnostic?	What does it measure or possibly detect?
Urine test for mom		
Blood test for mom		
Blood test for fetus		
Amniocentesis		
Chorionic villus sampling		
APGAR test on newborn		
Blood test on newborn		

6.7.3 Screening and diagnostic tests can be helpful, but they also raise difficult questions. Explain.

Learning Objective 6.8: What tests are on the horizon?

> *Evaluate the ramifications of further refinements in genetic tests and treatments.*

6.8 Summary

Scientists continue to work on screening tests that are less invasive and more accurate. Fetal DNA circulates in the mother's bloodstream, and this can be collected and copied to look for genetic disorders. Gene therapy, the correction of a defective gene, may be possible for embryos. But these tests are likely to be expensive and therefore not available to everyone, and how do we decide what "healthy" really means?

6.8 Define this Key Term

gene therapy

6.8 Questions for Review

6.8.1 What could gene therapy do for an embryo?

6.8.2 If there were expensive tests and procedures that could identify more embryonic "defects" than we can find now, would that be a good thing? Why or why not?

In every chapter in your text, the following five sections appear as special boxes within the chapter. You should refer to those as you try to answer the review questions here, which focus on the main point(s) of the section.

6. Biology in Perspective

Because it ensures the survival of our species, reproduction is one of the most important things we do. The first thing we ask on hearing about the birth of a baby is whether it is a boy or girl. Most of the time, eggs and sperm develop fine, fertilization occurs, hormones line up, and the baby is healthy. But not always. Yet a disorder of sexual development is somehow seen as profoundly worse than a circulatory system defect or heart problem. We expect sex and gender to be very well defined and "correct." It is not always that way.

Question to consider:

Why do you think developmental problems affecting a person's sex or gender seem so much more upsetting than problems with other body systems?

6. Scientist Spotlight: Dr. Virginia Apgar (1909–1974)

In the early 20th century, mothers were generally under general anesthesia when their babies were born, and the babies were whisked off to the nursery without an examination. That works fine for 90% of babies, but about 9% need some medical attention, and about 1% may not survive without it. Dr. Virginia Apgar was an anesthesiologist who assisted with thousands of deliveries and realized there was a real need to identify newborns that needed help.

Apgar studied how anesthesia given to mothers affected their babies and designed her test in 1952 as a guide to determine whether a baby needed resuscitation. Now know as the APGAR test, it's been used ever since. Apgar scores taken within 5 minutes of birth accurately predict both newborn survival and neurological development.

Questions to consider:

Why did Dr. Apgar go into anesthesiology rather than surgery?

What good is the APGAR test?

6. Technology Connection: Home Pregnancy Tests

Before 1978, the only way to find out whether a woman was pregnant early on was to see a doctor. Pregnancy tests then became available. These tests all work the same way. An antibody detects hCG, a hormone only present when there is a pregnancy.

The test is 99% accurate.

Questions to consider:

Why might a woman not want to see a doctor to find out if she was pregnant, and what's the problem with waiting?

How does a home pregnancy test work?

Why is it a good idea to test twice, a few days apart?

6. Life Application: Gender Testing in Sports

As far back as the 1936 Olympics, gender of successful women athletes was questioned because they were so successful in their sports. By the 1966 European Athletics Championship, women had to agree to a humiliating examination by a gynecologist and stand naked in front of a committee to confirm their external genitals to be allowed to compete. Chromosome testing to verify gender began with the 1968 Olympics. By 1991, a test for the SRY gene (that controls development of testes) was used to confirm gender. No one has ever been found to be deliberately misrepresenting her gender, but lives have been ruined.

There is no evidence that women athletes with DSDs have an edge. Testing was more driven by the fear that men would masquerade as women to win competitions. In 1999, the Olympic Committee finally abandoned compulsory gender testing.

Questions to consider:

Why did people think it was important to confirm gender for athletic competitions?

Why is it difficult to confirm gender?

6. How Do We Know?: The Female Reproductive Tract Actively Helps Sperm Find an Egg

Sperm has a long way to swim to find an egg. It gets help from the female reproductive tract, and in the process, the best sperm gets selected. After ejaculation into the vagina, sperm first face cervical mucus. Prior to ovulation, the molecules in the mucus are arranged like a mesh, catching the weak swimmers. As soon as a fertile egg is available, the fibers line up, allowing the sperm through. Only the healthiest sperm are then capable of binding to the oviduct to await ovulation. Once ovulation occurs, the sperm are set free. Fertilization can happen as many as 5 days after intercourse, so the sperm can hang out in the oviduct until the egg comes along. The egg releases chemical cues that help the sperm find it, and only the healthiest sperm "win" and fertilize an egg.

Questions to consider:

How does the female reproductive tract ensure that fertilization will only happen from a healthy, robust sperm?

Why does that matter?

Student Study Guide

Chapter 7 Plants, Agriculture, and Genetic Engineering

Can We Create Better Plants and Animals?

For billions of years, as a result of mutations and natural selection, each generation is genetically different than the previous one. This may be viewed as a naturally occurring form of genetic engineering. Humans have intervened in domesticated animals and crops, selecting and breeding for the most desirable traits. This process selects for the genes that result in those traits, so this is genetic engineering too. In modern genetic engineering though, we can, in theory, take genes from almost any species and insert them into almost any other species. This is one reason it is important to protect the biodiversity, or living things, on the planet: they represent a treasure trove of genes to work with and one an organism disappears, so do its genes.

Is genetic engineering safe, though, and is it wise?

7. Case Study: Golden Rice

Summary

Many children in the world have poor diets and suffer from vitamin A deficiency (VAD). VAD can damage eyes, leading to blindness, and can increase susceptibility to infections, which kills about half these children. Most of the cases of VAD occur where rice is the major source of calories, and otherwise there isn't access to balanced nutrition, as vitamin A is found in carrots, tomatoes, meat, and milk, for example. Scientists realized that fortifying rice with vitamin A would reduce malnourishment and save lives. At the time, the only way to do this was to take genes from a species that could make vitamin A and insert them into rice. Not a simple task.

A decade later, a Swiss scientist who was an expert in manipulating genes in rice, Potrykus, and a German scientist who was a specialist in the biochemistry of vitamin A, Beyer, happened to be traveling to New York City to a conference organized by the Rockefeller Center. They realized their specialties put them in a unique position to solve the problem of making rice a source of vitamin A. They were funded by the Rockefeller Foundation, and in 1999, *golden rice* was created.

In this process, much was learned to advance the field of genetic engineering, but complex questions have been raised about safety and ethics. What should we do with these new creative powers?

Case Study Questions for Review

Why did scientists choose to try and combine vitamin A with rice?

What is *golden rice*?

Learning Objective 7.1: What are the goals of genetic engineering in plants?

Describe the goals for genetic engineering of crop plants?

7.1   Summary

Golden rice is a genetically modified organism (GMO). The intent was that the rice seeds be distributed, as a humanitarian effort, to low-income farmers worldwide for just the cost of the seeds. Many corporations, however, see huge profits in genetically engineered crops, and this is big business, although these are banned in Europe and Japan because of popular opinion and safety fears. Typically, three goals motivate genetic engineering of crops:

1. Pesticide Production

If crops could produce their own toxins to kill pests, farmers wouldn't have to spend billions of dollars to spray pesticides, and pests wouldn't eat the plants. A bacterium named *Bacillus thruingiensis* or Bt, produces toxins that can kill certain caterpillars, beetles and fly larvae, which eat leaves and can devastate entire fields. It does not appear to harm helpful pollinating insects or other predators, nor people. The Bt toxin is a protein, so there are genes that code for it. The toxin gene can be extracted from the bacterium and inserted into the cells of a young crop plant, which will then make its own pesticide.

2. Herbicide Resistance

Weeds are even more trouble than insects, and killing weeds without killing the plant you're trying to grow is tricky. Building resistance to weed killer, or herbicide, in a crop plant is possible with genetic engineering. Glyphosate is the most widely used herbicide in the United States, sold under the name Roundup. There is an enzyme in a bacterium named Agrobacterium which is not affected by glyphosate. Genetic engineers take the gene that makes this enzyme from the bacterium and insert it into crop tissues, making the crops resistant to Roundup. They are called "Roundup Ready" and will survive being sprayed with Roundup, which kills all the weeds.

3. Increased Nutritional Value

Almost all genetically engineered crops grown in the United States have been engineered for pesticide production, herbicide resistance, or both. But in some cases, the goal is to enhance the nutritional value of a food, as with the golden rice. For that, two genes that caused the seeds to produce a molecule called B-carotene were inserted into the rice. B-carotene is found in fruits and vegetables that are orange colored as a result and gives the rice its golden color. Our bodies produce an enzyme that converts B-carotene to Vitamin A. Here are a few other examples:

- Soybeans and canola have been engineered to produce omega-3 fatty acids, thought to reduce risk of heart attach and stroke.
- Virus-resistant squashes are available.
- Sugar cane has been engineered to produce more sugar.
- A strain of potato makes a more edible form of starch.
- Drought- and freeze-resistant crops have been engineered.

Some of these foods have achieved commercial success, but sometimes they don't work. For example, tomatoes were engineered to ripen slowly so they would be easier to transport after harvesting, but they didn't taste good.

7.1 Define this Key Term

genetically modified organism

7.1 Questions for Review

7.1.1 Fill in the following table:

Goal of genetic engineering	Describe how this engineering is helpful	Give an example
Pesticide Production		
Herbicide Resistance		
Increase Nutrition		

7.1.2 Which crops in the United States are typically genetically modified?

7.1.3. Even if you don't think you eat any of the GM foods, you probably actually do. In what other foods are they often used?

7.1.4. How does a chemical company like Monsanto make money with GMOs?

Learning Objective 7.2: Why are plants such good sources of food?

> ***Explain how the bodies of plants are structured and specialized to store food molecules.***

7.2 Summary

Plants are good sources of food for us for two reasons:

1. By photosynthesizing, plants can capture solar energy and turn it into energy-containing chemical compounds that our cells can use for energy: food. They do this for themselves, but we can use many of the same energy-containing compounds they can.

2. Plants use some of the food they produce right away, but they store a lot of it in their roots, stems, leaves, and seeds. This is good for us because it gives us food in "packages" we can harvest.

Most of the plants we're talking about here are the flowering plants, the most common type of plant now living on Earth. They have four types of organs: roots, stems (branches), leaves, and flowers which produce seeds and fruit. These are connected by two vascular, or internal transport, systems, similar to your circulation system: xylem and phloem. All the plant organs can be used as storage for vitamins and edible compounds we can use.

Roots

Roots anchor the plant to the ground and absorb water and nutrients from the soil. Usually, they're underground and grow downward, away from the light. The younger, growing tips of the roots produce a dense array of tiny root hairs, which provide a large surface area for easy absorption of water and minerals.

Many plants store food molecules, especially starches, in their roots. This is because the part of the plant above ground may die over the winter, but the energy in the root will stay there for regrowth in the spring, or to provide energy for flowers and seeds. We eat a lot of roots: carrots, radishes, rutabagas, beets, parsnips, turnips, sweet potatoes. (Note that regular white potatoes are NOT on this list.)

Not all roots start at the ground. In the rainforest, plants like strangler figs start as seeds high in the trees and send their aerial roots down to the soil. Also in the rainforest, dirt and organic matter accumulates in the upper branches of trees, forming a mat of soil. The trees may grow roots from their upper branches to penetrate these aerial mats.

Stems

Stems connect the roots to the leaves and flowers. They provide structural support, allowing plants to grow taller to compete for sunlight. The vascular system within the stems and roots ties the plant together physiologically. Xylem transports water and minerals from roots to leaves and flowers. Phloem carries food made in the leaves through the stems to the roots and the rest of the plant.

Along the stem are nodes. Each one has a leaf and a bud. Some buds grow into stems that will become the plant's side branches, and some grow into flowers. There is also a bud at the tip of each stem which grows the stem longer.

The shape of a plant is determined by which buds grow and how long the stems get. Often the bud at the stem's tip, the terminal bud, prevents all the other buds on the stem from growing. This way the plant elongates toward the sunlight. If this bud is eaten or pinched off, the side buds grow and the plant gets bushier. Buds all over the plant ensure that if some part of it gets damaged, it can grow elsewhere.

Some stems are storage organs, like some roots. Potatoes, for example, are underground stems called tubers. Each "eye" is actually a node where a new stem and leaves could grow. Other stems are sources nutrients, vitamins, fiber, and other useful products.

Underground stems called rhizomes grow just below the surface to help a plant spread out. We see this in irises and ginger. In strawberries, horizontal stems called stolons grow just above the ground for the same purpose. Cactus pads are stems flattened to take on many of the functions of leaves.

Leaves

Leaves are typically broad, flat structures attached to stems by a stalk. Leaf shape is a good way to identify plants. Leaves are the plant's biochemical factories where photosynthesis occurs and food molecules are made, so they tend to be quite nutritious.

Sometimes leaves can be storage organs. Bulbs are short, underground stems with leaves that store nutrients and food. Onions and garlic are these, and their spiciness keeps herbivores from eating them. Succulent plants, which grow where it is dry, store water in their leaves. Aloe, ice plants, and stone plants do this.

We mostly eat regular aboveground leaves for their nutrients, vitamins, and fibers. "Leafy vegetables" include lettuces, spinach, and many others. The herbs we use for flavor are often leaves.

In a cactus, stems take on the role of leaves and the leaves are modified spines. In climbing plants, like ivies, all or part of the leaf is modified into a tendril to anchor the plant as it climbs. Carnivorous plants, like Venus flytraps, sundews, pitcher plants, and butterwort, have leaves adapted to catch and digest insects.

Flowers

Flowers come in many colors, sizes, shapes, and smells, each unique to its species of plants. Their purpose is sexual reproduction.

Like other living things, plants spend considerable energy on reproducing. This energy is directed toward two purposes: providing offspring with a food supply as part of the seed, and providing animals with a food supply to entice them to disperse young plants to new places (in fruits).

Flowers are highly modified leaves on a short stem. Four whorls of leaves each do something different:

1. Sepals: outermost whorl of leaves, which often form the outer protective layer on a flower bud

2. Petals: often the showiest part designed to attract pollinators

3. Stamens: male parts of the flower

4. Carpels: female parts of the flower

Flowers are edible (broccoli and cauliflower), but their main nutritional value comes from seeds and fruit. Seeds and fruits form a large portion of the diet of virtually the entire human population.

Seeds

A seed consists of an embryonic plant with a root, stem, and one or two leaves. The endosperm are cells making up most of the seed, surrounding the embryo. The endosperm is full of sugars, starch, proteins, amino acids, and other nutrients that feed the embryo and seedling when it germinates (begins to grow) until it is big enough to make its own food. A seed coat covers the whole thing to keep it from drying out.

We can use the endosperm for food too. Nutrition from corn kernels, beans, rice, wheat, barley, peas, peanuts and many other seeds comes from the endosperm. We also get corn, cotton, safflower, and peanut oil from seeds.

Fruits

Seeds form inside the female part of the flower in the ovary, and as the seeds ripen, the ovary develops into a fruit. Any structure that envelopes the ovary is, botanically speaking, a fruit. So tomatoes and those little tufts on dandelion seeds are both included. The purpose of a fruit is to help the seeds disperse because they and their parents can't move around on their own.

Fruits we eat are created by the plant to entice animals to eat them as a way to disperse seeds. Seeds have that protective coat, so they pass through digestive systems unharmed and get deposited in a new place, even inside a bit of "fertilizer."

7.2 Define these Key Terms

xylem

phloem

roots

root hairs

stems

buds

leaves

flowers

fruits

seeds

tubers

rhizomes

stolons

leaves

flowers

sepals

petals

stamen

carpals

7.2 Question for Review

7.2.1. List examples of the following:

      a. A stem we use for food

      b. A root we use for food

      c. A leaf we use for food

      d. A seed we use for food

      e. A flower we use for food

Learning Objective 7.3. How do plants make food?

Summarize the light reactions and Calvin cycle phases of photosynthesis.

7.3 Summary

Plants use a process called photosynthesis to convert solar energy, carbon dioxide, and water to sugar. Directly or indirectly, all the energy used by most species on the planet comes from sunlight captured by photosynthesis. Oxygen is released in the process.

$$\text{Water + carbon dioxide} \xrightarrow{\text{sunlight}} \text{sugar + oxygen}$$

Sugar is a carbohydrate, or carbon bonded to water: CH_2O. The formula tells us that in a carbohydrate there is a ratio of 1 Carbon to 2 Hydrogens to 1 Oxygen. The specific sugar made by photosynthesis is glucose which is $C_6H_{12}O_6$. To get to $C_6H_{12}O_6$, we need six molecules of water and six molecules of carbon dioxide, so the equation is

$$6H_2O + 6CO_2 \xrightarrow{\text{sunlight}} C_6H_{12}O_6 + 6O_2$$

Sunlight provides the energy for this, and some of that energy is stored in the bonds of the glucose. Plants are able to store 1–5% of the available solar energy as sugar.

Light Reactions and the Calvin Cycle

There are two parts to photosynthesis. The light reactions of photosynthesis capture solar energy, and the Calvin cycle uses the energy to make sugar.

A green pigment called chlorophyll is found in all green parts of a plant. In the light reactions, when a molecule of chlorophyll absorbs sunlight, the solar energy is transferred to an electron, which shoots out of the chlorophyll. This electron is transferred to a series of other compounds, releasing a little energy, or ATP, with each transfer, and ends up in NADPH. The chlorophyll can't be left with "holes" where it lost its electrons, so water molecules are also split during the light reactions, releasing the hydrogen to fill in the holes, and oxygen is released into the atmosphere. NADPH and ATP store chemical energy and the Calvin cycle uses that energy to make sugar.

Because the Calvin cycle gets its energy from NADPH and ATP, it doesn't require sunlight. Carbon dioxide from the atmosphere is attached to another carbon-containing compound—

carbon fixation. In a series of chemical steps, energy from the NADPH and ATP is added to this developing molecule, which is eventually the energy-rich molecule glucose. Cells use glucose for energy directly, or they convert into many other compounds they need.

A photosynthetic cell contains organelles called chloroplasts in which photosynthesis occurs. The light reactions occur in the green portions of these chloroplasts.

Leaves

Because photosynthesis requires water, carbon dioxide, and sunlight, the main function of a leaf is to provide all this. The shape of the leaf promotes light capture because most are broad and flat, placing the cells close to the surface so the sun reaches them all. Inside the leaf, the cells are loosely packed and spongy to allow carbon dioxide to circulate. Water enters the leaves through the circulatory system of the plant, where it evaporates into the inner part of the leaf and then diffuses into the photosynthesizing cells. Humidity inside the leaf is close to 100%.

The outer layer of cells on a leaf are tightly packed and secrete wax. This makes them waterproof and prevents dehydration by evaporation. Small pores called stomata in the leaf surface allow diffusion of carbon dioxide and oxygen. The stomata can close to conserve water, although while they are closed, the plant cannot photosynthesize.

7.3 Define these Key Terms

photosynthesis

glucose

light reactions

chlorophyll

carbon fixation

chloroplasts

stomata

7.3 Questions for Review

7.3.1 Without looking it up again, write the chemical reaction for photosynthesis.

7.3.2 Why does chlorophyll look green? What wavelengths of light does chlorophyll absorb (and therefore get its energy from)?

7.3.3 Draw a simple flow chart of the steps in the light reactions and Calvin Cycle. Indicate where energy is taken up and released.

7.3.4 What are chloroplasts? Where are they and what do they do?

7.3.5 How does a leaf help a plant photosynthesize?

7.3.6 How does a leaf conserve water?

7.3.7 Why do stomata often close up at night? (Why close up at all, and why at night rather than during the day?)

Learning Objective 7.4: How was golden rice engineered?

Outline the steps used to produce a genetically modified organism

7.4 Summary

Each genetic engineering project tries to insert working copies of a gene from one organism into another. Because each organism has different genes and different organisms, each project faces unique challenges, but they follow the same basic steps, which we'll illustrate with the golden rice project:

1. Define the problem.

Plants need vitamin A, and most plants can make their own. Rice plants make vitamin A in their leaves and other organs, but not in the seeds, which is what we eat. The scientists Potrykus and Beyer discovered that the seed endosperm cells were missing two enzymes necessary to make vitamin A. It turned out that they had the genes to make the enzymes, but they were inactivated. So the problem was this: how can we insert working copies of these enzyme genes into endosperm cells of rice?

2. Clone the genes.

It would make the most sense to use the rice genes for the missing enzymes, but Potrykus and Beyer couldn't figure out how to activate them again, so they used genes from two other plants that have the same enzymes that make vitamin A: daffodils and a soil bacterium. The next step was to biochemically clip out the relevant genes from the rest of the plant chromosomes, and then make billions of copies of each to have enough to work with. This isolating and copying process is called cloning.

3. Package the genes.

Simply inserting a gene into an organism doesn't usually work either because the organism's enzymes break down the foreign DNA, or the correct biochemical network is absent. The gene must be protected and processed to be part of a DNA package the cell can use. To do this, extra DNA is added at the beginning of the gene to show where the gene starts and to provide a target for enzymes to start transcription. A "stop" section is then added at the end of the gene to tell the enzymes to stop transcription. In the case of the golden rice, they also needed to add sections of DNA to ensure that the genes only work in the endosperm of the rice, and to add markers to indicate successful gene insertion.

4. Transform the cells.

When a cell takes in and uses foreign DNA, the process is called transformation. There are a few ways to do this; Portrykus and Beyer selected the method that contributed to the Roundup Ready gene. The bacterium *Agrobacterium* is first transformed by the addition of the enzyme genes from the other plants. Bacteria transform easily by taking up foreign DNA. Pieces of rice plant were bathed in a solution of transformed *Agrobacterium,* and many of the fragments became infected, successfully absorbing the DNA packages.

5. Confirm the strain.

The transformed pieces of plant were placed in a growth medium, grew up to whole plants, and produced seeds. In a few of these cases, the plants had incorporated the foreign genes, and the endosperm cells in the seeds produced vitamin A. These plants could then be interbred; their offspring then produce vitamin A; and we have golden rice.

7.4 Define this Key Term

cloning

7.4 Questions for Review

7.4.1 Draw a flow chart indicating the five steps required to genetically engineer an organism. Describe what happens at each step and why it is necessary.

Learning Objective 7.5: How else is genetic engineering being used?

Describe how areas other than agriculture use genetic engineering.

7.5 Summary

Humans have been selecting for desirable traits in crops and animals since we've been farming. Genetic engineering is now being used in other areas, too.

1. Medicine

Historically, diabetes has been treated with insulin from pigs or cattle because human insulin was expensive and not available in large quantities. In 1978, a scientist successfully inserted the human insulin gene into a bacterium, which then produced large quantities of human insulin. Today, most insulin used is human insulin produced by bacteria, yeast, or other organisms.

Bacteria or yeast are usually used to produce drugs like these:

- o Human growth hormone, used to treat dwarfism
- o Blood clotting factors, used to treat hemophilia
- o Erythropoietin, used to treat anemia
- o Vaccines like Gardasil, which helps prevent cervical cancer
- o Human antithrombin, in which a gene inserted into a goat produces the protein in its milk which can then be extracted and purified as a drug to reduce blood clots

2. Industry

Bioremediation is the biological cleanup of pollutants or contaminants. Organisms can be genetically engineered to break down those chemicals. For example, bacteria have been engineered to break down oil spills.

Some bacteria can survive in harsh environments with high acidity or salinity, or low oxygen. Bacteria have been engineered to produce enzymes that digest metals from their ores, making them easy to extract. This process is called bioleaching. These can also be used in cleanups of heavy metals.

Bacteria have been genetically engineered to make materials used in the production of plastics, textiles, and cheese. Scientists are working on developing strains which can digest waste crop materials or wood, and excrete compounds similar to crude oil.

3. Research

There are several varieties of genetically engineered mice used by researchers. One type is highly susceptible to cancer and provides a large population on which to try new treatments. Others called "knockouts" have had specific genes removed, allowing scientists to discover what those genes do.

4. Novelty

Genetic engineering has resulted in a few interesting commercial products:

- o   Glofish: modified zebrafish that glow green, red, or orange
- o   Blue rose: modified with a gene from a petunia
- o   Bioart: bacteria that have had genes for fluorescent proteins inserted so they can produce colorful displays

5. Construction of Organism's Genomes

A genome is all of an organism's DNA, with every gene required to make it work. In theory, if we removed DNA from a cell and inserted an entire synthetic genome, we could make some new form of living thing. In May 2010, scientists announced that they had synthesized an entire genome of one species of bacteria, inserted it into a cell, and that the cells grew and reproduced normally: a line of living cells without a parent. This did not involve creating new genes though, since they were copied. But genes from different organisms could be mixed and matched in new ways to create new, artificially created forms of life.

7.5 Define these Key Terms

bioremediation

bioleaching

genome

7.5 Questions to Review

7.5.1 Fill in the following table:

	How genetic engineering is used	Examples (list at least two for each)
Medicine		
Industry		
Research		
Commercial/Novelty		

7.5.2 Can scientists create artificial life through genetic engineering?

Learning Objective 7.6: What are the risks of genetic engineering?

> *Assess the risks of genetic engineering in terms of safety, the economy, and effectiveness.*

7.6 Summary

Custom-designed life forms have the potential to produce huge changes in the way we live, promising large benefits, but also risks. Golden rice provides an example of how the debates over this play out.

1. Safety

The B-carotene produced in golden rice seeds is chemically identical to that found naturally in other plants. Might there be some unanticipated harmful reaction caused by the inserted gene? Could these enzymes be making harmful products along with the vitamin A? Does the B-carotene get converted into something else that is potentially harmful?

Genes make proteins, and proteins can trigger allergic reactions. Genetic engineering combines genes from several species into one, which also combines potential allergens. Odds are against this because, of the tens of thousands of proteins produced by a species, genetic engineering uses only a couple. But some people are allergic to something in daffodils, so might they also be allergic to golden rice? Is the current rise in allergies related to increased use of genetically engineered food?

Critics of genetically engineered food refer to it as "Frankenfood," cobbled together like Frankenstein's monster. Some people want to see "natural" on their food labels. Usually, genetically engineered foods are not labeled as such. Genetically engineered foods also tend to be used in processed and fast foods—high fructose corn syrup, soybeans.

Scientists and researchers are aware of these issues and feel that they have tested the golden rice thoroughly. No safety issues have been identified.

2. Economic Considerations

It takes a large corporate investment to produce a genetically engineered crop, and corporations expect to make a profit. More than 70 patents came out of the process that resulted in golden rice. Syngenta owns most of these, but more than 30 other corporations are involved. Potrykus, Beyer, and Syngenta worked with the Golden Rice Humanitarian Board to persuade each of these corporations to forgo licensing fees for low-income farmers. Farmers can also keep the seed from one year to the next. Syngenta could profit from golden rice in developed countries, but those people aren't deficient in vitamin A. The company has given up commercial interest in golden rice, but most companies that invest in this technology guard their patents fiercely and have sued each other and even farmers, over patent infringement.

3. Effectiveness

Is golden rice a good solution for vitamin A deficiency, or would it be better to invest money in finding ways to give these people a more balanced diet? No doubt, it's been a significant scientific achievement, but it is still not commercially available, and there is still debate on the benefits versus the risks.

(There were no Key Terms for this section.)

7.6 Questions for Review

7.6.1 Genetic engineering is somewhat controversial in the areas of safety, economics, and effectiveness. Explain the controversy in each of these cases, using golden rice as the example.

Learning Objective 7.7: How ethical is genetic engineering?

> **_Evaluate the ethical considerations that confront scientists related to genetic engineering._**

7.7 Summary

Genetic engineering raises several ethical issues, generally related to environmental impacts and impacts on humans.

1. Genetic Engineering and the Environment

Genes in nature are normally transferred from parent to offspring: vertical gene transfer. Genetic engineering transfers genes between individuals in the same generation, often between different species: horizontal gene transfer. Bacteria normally use horizontal gene transfer, so if genetically engineered bacteria got out of the lab environment and into nature, their engineered genes could be incorporated into other bacteria.

Plants can also engage in horizontal gene transfer and can hybridize. Pollen from a genetically engineered crop can easily get out of a particular field and possibly introduce engineered genes into other plants. For instance, what if Roundup Ready corn hybridized with a grassy weed and the weed, therefore, became resistant to Roundup?

These examples are sort of accidental, but what if someone meant harm? A genetically engineered virus that was as toxic as anthrax, as virulent as Ebola, and as contagious as the flu?

2. Genetic Engineering and Human Life

What if we could create life forms to our own specifications? Do we have the right to genetically alter species to suit our needs? Do we have a responsibility to do this to improve the food supply for the world?

Should we genetically engineer ourselves? Maybe we could get longer lives, better health, cure diseases, chose greater intelligence, athletic ability, or beauty. We can't do these things yet, but the basic techniques are there. Should we do these things?

7.7 Questions for Review

7.7.1 Explain an example where genetic engineering using bacteria could be both beneficial and risky.

7.7.2 Explain a few ways in which we could, in theory, genetically engineer humans.

In every chapter in your text, the following five sections appear as special boxes within the chapter. You should refer to those as you try to answer the review questions here, which focus on the main point(s) of the section.

7. Biology in Perspective

Since we domesticated plants and animals, we have been selecting for traits beneficial or desirable to us. Genetic engineering allows us to do this with more precision, much more quickly, and we can take traits from one species and put them in another. There are very real concerns about economics, safety, and the environment, and ethical concerns, especially related to engineering human genes.

Question to consider:

What can genetic engineering do that ordinary selective breeding of crops and animals could not?

7. Scientist Spotlight: Kary Mullis (1944–)

Kary Mullis is famous for developing the now common technique of copying DNA segments. It is called the polymerase chain reaction or PCR and is the first step in every genetic engineering project. PCR clones the gene so there are enough copies to work with. It's used in genetic engineering, paternity testing (comparing a possible father's DNA to a child's), and examining DNA in criminal cases. PCR is a cornerstone of modern molecular biology, and in 1993 Mullis and his colleague Michael Smith won the Nobel Prize in chemistry for this work.

Question to consider:

Trace the steps in PCR. Why is it a useful tool?

7. Technology Connection: How to Transform Cells

A cell is transformed when it takes up foreign DNA from the environment and that DNA becomes a working part of the cell's genetic material. In a lab setting, it is easy to force bacteria to take up DNA. That's why bacteria are commonly used in genetic engineering. It is more difficult to transform plant and animal cells, but there are several techniques for doing this. Once the DNA is in the cell, in plant and animal cells, foreign DNA would get digested, so it must be incorporated into the chromosome.

Question to consider:

How can a bacterial cell become transformed?

How can a plant or animal cell become transformed?

7. Life Application: From Teosinte to Maize

Some of the earliest genetic engineers lived about 9000 years ago in what is now southern Mexico. They converted a native plant, teosinte, into maize, or what we know as corn. The differences are due to only five genes. Over years of harvesting the plants with the traits found in modern corn and planting the seeds from these plants, teosinte evolved into maize, with its few ears with many big, soft, kernels, that stay on the cob.

Question to consider:

Explain how early people harvesting teosinte could have driven the shift toward what we now know as modern corn via their selection of plants.

7. How Do We Know?: Are Genetically Engineered Products Safe?

There are several ways to look at this question.

First, there is the principal of substantial equivalence, which says that a genetically engineered product is safe as long as it is "substantially equivalent" to the non-engineered product. Critics of genetically engineered products argue that this isn't enough, that tests should be done to see what happens when humans use these things. It's also a problem that the developers of the products are also responsible for testing the products. The best bet is to stay informed and make decisions for ourselves on these issues.

Question to consider:

What is the principal of substantial equivalence, and how does it apply to genetically engineered products?

Student Study Guide

Chapter 8 Health Care and the Human Genome

How Will We Use Our New Medical and Genetic Skills?

We now know things enough about inheritance and genetics that we have the capacity to do wonderful things in disease prevention and cure, but we also have the capacity to do mischief and harm. We will have to decide the ethics of what we should do *versus* what we can do.

8. Case Study: Carrie Buck and the American Eugenics Movement

8. Case Study Summary

Emma Buck was poor and uneducated, had been arrested for prostitution and suffered from syphilis, and in her rural Virginia community in the early 1920s, she was brought before the Commission on Feeblemindedness. The Commission declared her "suspected of being feebleminded or epileptic" and committed her to the Colony for Epileptics and Feebleminded for the rest of her life. Emma's daughter, Carrie, was taken in by the Dobb's family at age 14.

The Dobb's were good-hearted people, and Carrie did well in school and did chores around the house. But in 6th grade, the Dobbs' took her out of school to focus on the housework. In 1923, Carrie was 17 and became pregnant. She said she was raped, but in that era, any pregnancy outside of marriage was deeply shameful, so the Dobb's immediately filed paperwork to have Carrie committed like her mother.

In between Emma's and Carrie's commitments, Virginia passed a Eugenics Law which said that to improve the "human stock" it was necessary to prevent biologically defective people from having children. People should be sterilized for "the good of society." Feeblemindedness was considered hereditary, so Carrie's daughter Vivian was suspected of feeblemindedness. A social worker declared her "not quite normal," so she was labeled as feebleminded.

The case Buck v. Bell challenging the Virginia law went to the Supreme Court in 1927 and the law was upheld. Carrie was sterilized in 1927 at age 21. Vivian died of an infection at age 8, but had made the honor roll in school so clearly wasn't "feebleminded."

Is eugenics simply a shameful chapter in our history, or is the goal of improving humans still alive with us today?

Questions for Review:

What is eugenics?

Who was Carrie Buck?

Learning Objective 1: Do Complex Human Characteristics Have a Genetic Basis?

Assess the legitimacy and risks of genetic determinism to society

8.1   Summary

Eugenics sterilization laws prevalent in the United States in the early 20th century said that a board of experts could have a person sterilized if they were judged to be "insane, idiotic, imbecilic, or moronic," or were convicted criminals. In some cases alcoholism, drug addiction, blindness, deafness, homelessness, or poverty could also get an individual sterilized, all on the grounds that these defects were heritable, so the only way to remove them from the population was by sterilization.

Genetic essentialism says that being human means having a human genome, the complete set of genes for a particular organism. All of our cells, in fact, have a human genome, but that is certainly not the only thing that makes us human.

Genetic determinism is the idea that our genes determine or direct everything about us: physical appearances, emotions, and behaviors.

We know, though, that identical twins, even conjoined identical twins, have distinct personalities, likes, dislikes, and behaviors. Our genome defines our biological potential: it makes us humans, but it is responsive to the environment even before we are born.

There are no specific genes that define human abilities or personalities. Genes play a role but do not determine the outcome. A person with a gene for alcoholism would not develop the disease if he never drank alcohol.

Eugenics tries to define "normal." But normal can be a statistical average, or it can be whatever is closest to society's ideal, for beauty, for example. Variation is normal, however. Most individuals in a population deviate from the average, and what we view as normal varies. People used to kill redheads because they were thought to be witches.

Trying to define normal as opposed to diseased or disabled is sometimes equally difficult. Dwarfs are short, but otherwise do not necessarily have health problems. Breast cancer is a disease, not just a variation in a trait.

It's still difficult to define normal, but today we have genetic screening, prenatal diagnosis, and gene therapy, all designed to hunt down and fix "bad" genes.

8.1 Define these Key Terms

genetic essentialism

genetic determinism

8.1 Questions for Review

8.1.1 What was the goal of eugenics, and what are a few of the "traits" which would label an individual as bad for society?

8.1.2. Explain what genetic determinism is with an example.

8.1.3. List some examples of evidence that the genes we have is not what determines who we are.

8.1.4 List some examples illustrating why it is hard to define "normal" in humans.

8.1.5 We no longer have Commissions on Feeblemindedness, but do we still try to eliminate "bad" genes? How do we do it now?

Learning Objective 8.2: What is Gene Therapy?

Distinguish between somatic and germ-line gene therapy.

8.2 Summary

Thousands of diseases and disorders can be attributed to mutations in specific genes. There are many examples in the textbook. About one half to two thirds of all conceptions result in miscarriages caused by serious mutations.

Genetic screening of embryos and fetuses allows us to identify genetic or developmental abnormalities, and parents use this to prepare for a special needs child or to terminate a pregnancy because they don't want to pass on an unwanted trait. How do we decide which traits are acceptable and which we should seek to eliminate? In addition, not all genetic syndromes are predictable, and some mutations are new and not from a parent.

The goal of gene therapy is to cure genetic diseases. There are two types of gene therapy:

1. Somatic gene therapy

In this case, the idea is that a genetic disorder can be cured by inserting a normally functioning gene into the patient.

The first successful gene therapy happened in 1990, with children suffering from combined immunodeficiency (SCID). They are missing an enzyme (ADA), which is necessary for normal lymphocyte function. Lymphocytes are white blood cells; they are an essential part of the immune response, or the ability to fight off an infection. Immature lymphocytes and lymphocyte stem cells from SCID patients were cultured in the lab. ADA genes were isolated form healthy cells and inserted into harmless viruses. These then infected the cultured cells, and the ADA genes inserted themselves into the cells' genome. The cells now producing ADA were inserted back into the patient, and in successful cases, the new gene produces the missing enzyme, treating the disease.

2. Germ-line gene therapy

This therapy aims to correct genetic problems in the germ-line, or the eggs and sperm, so that harmful mutations cannot be passed on. This would, in theory, improve the "human stock" by eliminating bad genes. It has not been tried in humans yet, but it's routinely done in research with laboratory mice. The rhesus monkey ADNi has been altered so its DNA and RNA encode for a green fluorescent protein (GFP), although the GFP RNA isn't translated in this monkey.

The goal with the monkeys was to develop methods to insert human disease genes into primates so the disease could be studied in these animals which are so similar to humans. Is that ethical? Let alone engineering our own germ lines?

So far there are significant technical and scientific questions. The process of genetically engineering the germ-line is quite inefficient. The development of ANDi required hundreds of eggs out of which only three live births occurred. Scientists inserted GFP genes into harmless viruses and injected them into unfertilized rhesus monkey eggs. The eggs were fertilized with rhesus monkey sperm. The embryos were implanted into surrogate rhesus mothers which had undergone hormone treatments so they would accept the pregnancy. A few monkey surrogates carried the babies to term and three that were born incorporated the GFP into their genomes. Two of those were stillborn.

8.2 Define these Key Terms

somatic gene therapy

germ-line gene therapy

8.2 Questions for Review

8.2.1 Why do people have prenatal screening done? What are they trying to discover?

8.2.2 What is the goal of gene therapy in general?

8.2.3 Draw a flow chart illustrating the process of somatic gene therapy and draw one adjacent illustrating the process of germ-line gene therapy.

8.2.4. Contrast the two procedures? What are the differences in the process, and how are the results different?

Learning Objective 8.3: What are the Benefits and Risks of Genetically Altering Humans

Contrast the benefits and risks of somatic and germ-line gene therapy.

8.3 Summary

Somatic and germ-line gene therapy are quite different from each other. Somatic gene therapy fixes or replaces a defective gene in an individual. The fix is not passed on. In germ-line gene therapy, the genetic changes will be inherited by all descendants. This difference raises ethical challenges and different potential benefits and costs.

Somatic Gene Therapy

Somatic gene therapy could potentially cure genetic disorders. Why not fix these?

Some arguments: It's not safe, it might not work, we should not "play God."

New medical treatments are generally somewhat risky, and effectiveness will be variable. "Loss of function" diseases are easier to cure. For example, in hemophilia, blood does not clot because the gene associated with producing a clotting factor fails. Inserting a gene that did produce the clotting factor would cure the disease. It's more difficult to cure a disease caused by a gene that produces a "gain of function" like in sickle cell disease. In that case a mutant gene produces abnormal hemoglobin. Gene therapy would have to both insert a correctly functioning gene and disable or eliminate the mutant genes in all the blood cells.

Somatic gene therapy is expensive, but this is true of any new medical treatment. As for "playing God," one could argue we do that every time we try to treat any illness, disability or accident.

So far, somatic gene therapy has proven potential, but scientists don't have complete control over where the introduced genes insert and the consequences are potentially lethal, but it seems likely that development of this as a treatment will continue because it shows promise to help people.

Germ-Line Gene Therapy

Eugenics sterilizations attempted to ensure no "bad" genes were passed on by ensuring that no genes were passed on. Germ-line gene therapy is much more focused, but it does have two things in common with eugenics: (1) confidence about the deterministic genetic basis of characteristics and (2) confidence about defining desirable and undesirable traits.

There are problems with this. The relationships between individual genes and specific characteristics are not straightforward, and often a trait is influenced by multiple genes, or multiple genes influence one trait. For example, the globin gene affects both how well red blood cells carry oxygen and the shape of those cells. In addition, the environment may influence how a trait develops. Is this ethical?

We can avoid the issue with genetic counseling, which can inform couples about their risk of passing on specific genes associated with disease. Prenatal tests can determine whether an embryo is carrying the genes. With *in vitro* fertilization, only unaffected embryos can be implanted. Or there is egg or sperm donation.

Several ethical concerns arise with germ-line gene therapy. These are serious enough that many people think we should just stop trying to develop the technology to do it:

- The process is not necessarily safe, a mistake is permanent, and it will affect more than just one person. What if a different genetic function is accidently disrupted?
- We are not sure which genes are "good" and which are "bad." For example, having one allele for sickle cell disease is a good malaria preventative which is very helpful if one lives where malaria is common. It's just if one has two that it's trouble.
- We need to distinguish between medical treatment and enhancement. Is it acceptable to "design" children according to the parent's desires? What if fashion changes?
- This type of therapy could alter how we view the sick and disabled. Would it be frowned upon to not have perfect children? What if a family couldn't afford to get a genetic "defect" fixed?

8.3 Define this Key Term

genetic counseling

8.3 Questions for Review

8.3.1 What is the goal of somatic gene therapy?

8.3.2 Explain three arguments against gene therapy. Try to refute each one.

8.3.3 What is the goal of germ-line gene therapy?

8.3.4 How does genetic counseling maybe eliminate the need for germ-line gene therapy?

8.3.5 Explain four ethical issues related to germ-line gene therapy.

Learning Objective 8.4: How Can Stem Cells and Cloning Be Used to Alter People?

Compare the different uses of stem cells and cloning.

8.4 Summary

Like gene therapy, the use of stem cells and cloning also offers the hope of curing disease and the risk of unintended consequences.

Stem Cells

Stem cells have the ability for self-renewal. At each division, a stem cell can give rise to another stem cell as well as a more specialized cell.

The goal of stem cell research is that scientists will develop methods to use stem cells to replace damaged or diseased cells, like neurons in the spinal column damaged by injury or Parkinson's disease, or diseased pancreatic cells that could cure diabetes.

The concern about stem cell research centers on their source. Stem cells can come from early embryos, fetal cells, umbilical cord blood, babies, children, or adults. The stem cells from early embryos (the eight-cell stage) have the most developmental flexibility, or are totipotent, meaning capable of producing any of the cells in a developing embryo.

After the eight-cell stage of embryonic development, the cells split into the inner cell mass (ICM), which will develop into a baby. Those cells are the source for human embryonic stem cells. They can be isolated from 5–7-day-old embryos and grown in the lab and they are pluripotent. Pluripotent cells can produce any of the cells of the body but not the support structures needed for the development of a baby, so they can't produce a baby.

Human embryonic germ cells are collected from 6–9-week-old embryos and fetuses. These would ultimately produce the eggs or sperm, and they, too, are pluripotent. Stem cells collected later, from the umbilical cord, or a child or adult, have much more limited developmental potential.

Embryonic stem cells have the potential to treat Alzheimer's disease, heart disease, brain injury, blood cancers, immune system disease, and many others. But acquiring them destroys human embryos, and some people question whether this is ethical. Two questions are debated: When does human life begin, and what is appropriate to do with and to a human embryo?

Human embryonic stem cells can be collected from 5–7-day-old embryos that are composed of 100–150 cells at that point. One could argue that it's okay to generate human embryonic stem cells by combining donated eggs and sperm in the lab, just for this purpose. Or "extra" embryos from fertility clinics could be used. And some people believe none of this is acceptable.

The National Council of Churches has no official position. The text summarizes the views of various religions. Here are some facts:

- Human embryonic stem cells are isolated from embryos at such an early stage that no differentiation has occurred. The cells contain human genomes but are not persons.
- Human embryonic stem cells are not even potential persons because once isolated, they can't produce any of the tissues needed to support the development of a baby.

What do we do when religions come to different conclusions about issues that affect everyone? Should religion play a role in research decisions?

Cloning

Cloning can be divided into two categories: therapeutic and reproductive.

Therapeutic Cloning

Each of your cells has proteins on its surface that allow your body to recognize "self" and "non-self." This is why transplanted organs can be rejected; however, it is also how your immune system knows to attack foreign cells. Therapeutic cloning can solve this problem. In somatic cell nuclear transfer (SCNT), a nucleus is taken from a healthy cell in the patient and is placed into an egg that has had its nucleus removed. The transplanted nucleus directs the development of an embryo from which stem cells can be removed. The proteins on those cells will match the patient, eliminating transplantation rejection.

Reproductive Cloning

The goal here is to create embryos that will develop into offspring. As with SCNT, a nucleus from a "parent" cell is placed into an egg with its nucleus removed. The resulting embryo is implanted in a surrogate mother. The organism produced this way will be genetically identical to the parent that donated the nucleus. This has been successfully done with a variety of animals, but not humans. In more than 25 countries, cloning humans is illegal, although not in the United States.

8.4 Define these Key Terms

stem cells

totipotent

pluripotent

human embryonic stem cells

human embryonic germ cells

therapeutic cloning

reproductive cloning

8.4 Questions for Review

8.4.1 Why are stem cells potentially valuable for medical treatment? What are a few examples where they could be used?

8.4.2 Why are the stem cells from early embryos especially useful? What's different about them compared to adult stem cells?

8.4.3 What are three possible sources for embryonic stem cells?

8.4.4. Why is research with embryonic stem cells controversial?

8.4.5 Are all religions opposed to embryonic stem cell research? Examples?

8.4.6. Outline a fact-based argument in favor of embryonic stem cell research.

8.4.7 Diagram the steps in cloning.

8.4.8 Describe the difference in the goals and outcomes of therapeutic *versus* reproductive cloning.

Learning Objective 8.5: What are the Benefits and Risks of Stem Cell Research?

Use information on risks and benefits to evaluate the ethics of stem cell research.

8.5 Summary

Here are some predictions related to stem cell research:

- o Human embryonic stem cells may provide the key to alleviating suffering and curing disease; stem cells from the germ cells of fetuses also have significant medical potential.
- o Embryonic stem cells created by SCNT probably have the best chance to work well in patients because they're less likely to be rejected.
- o Umbilical cord blood and adult stem cells will be important tools in treating disease, but they are more limited than embryonic stem cells.

Benefits

Mouse and human embryonic stem cells have both been successfully directed to make specific cell types in culture.

Research with whole animals is promising. Human embryonic stem cells injected into rats suffering from the rat equivalent of Parkinson's disease differentiated into neurons. These produced dopamine and reduced the Parkinson's –like symptoms. Human embryonic stem cells that differentiated into heart cells were able to repair a heart defect in pigs.

Human embryonic stem cells teach us about normal human development and how birth defects occur. By altering specific genes in embryonic stem cells, it would be possible to create human disease models at the cellular level which could then be studied and experimented upon, rather than using mouse or other mammal models. Drugs could be directly tested on human cells.

Risks

Stem cell procedures are new and not completely safe yet. Transplanted stem cells are not always perfectly matched to patients and may be rejected. The dosage of stem cells must be carefully controlled, or tumors may form.

The donation of eggs for SCNT is also potentially dangerous, and a large number of eggs will be needed for research and medical treatments. To donate eggs, women must undergo hormone treatments so they will release multiple eggs. Between 1 and 2% of women who do this develop a severe form of ovarian hyperstimulation syndrome, which can result in kidney or lung failure, shock, or ruptured ovaries. It's possible that poor women would be exploited or pressured into selling eggs.

Stem cell and cloning technologies could be used for enhancement rather than for medical purposes. This could lead to human reproductive cloning.

Finally, these new technologies could extend human life beyond what is beneficial to society. Knowing one can always be "fixed" might lead to a less meaningful life, and an able-bodied older generation would not have to get out of the way of the next.

(There are no Key Terms specific to this section.)

8.5 Questions for Review

8.5.1 Explain the benefits of cloning and stem cell technologies.

8.5.2. Put together a fact-based argument against using cloning and stem cell technologies.

Learning Objective 8.6: What Other Challenges Result from Advances in Genetic Technology?

Assess the ethical challenges we face as a result of advances n genetic technology.

8.6 Summary

Versions of the eugenics law which resulted in the forced sterilization of Carrie Buck were enforced in 30 states by the 1930s. Forced sterilizations continued into the 1970s, in particular under a program of Native American Sterilization. We don't try to define "undesirable" quite as coarsely now, but we do make judgments about which traits and physical characteristics are good and which should be avoided.

Privacy

With the ability to "read" DNA sequences, we raise the possibility that information about risks we may carry or pass along to our children will be widely available. You might want to know this (or you might not) but what about your doctor, employer, insurance company, the police, or the government?

Accessibility

There are already wide disparities in the quality of health care people in the United States can pay for or even find, depending on where they live. These new technologies are likely to be expensive, so they will be beyond the reach of many. In fact, one could argue that it might be better to put our resources toward providing basic health care to everyone rather than spending money on these expensive, risky medical treatments that few may ever be able to access.

Danger of a New Eugenics Movement

Genetic screening makes sense to most people. The results provide parents with information about their child's potential future. If the embryo or fetus has a serious "problem," the parents can decide whether to continue with the pregnancy. They are deciding what traits are acceptable and what is not. In a survey conducted across 37 countries, 82% of geneticists in the United States supported the abortion of fetuses with Down syndrome, 92% supported the abortion of fetuses with spina bifida. The general public was not so supportive of termination of fetuses except in the case of extremely severe mental retardation.

Most people probably agree that prenatal screening for detection of fatal diseases is a good thing. But what if it's something like Huntington's disease, where the person may well live to be 50 years old? Is that not a life worth living? Should they have children?

Wealthier people who can afford this type of screening could end up having less risk of having disabled children than poor people, creating a world of genetic "haves" and "have nots."

8.6 Questions for Review

8.6.1 Eugenics is ancient history. We don't need to worry about that. Outline a fact-based argument refuting this.

8.6.2 How do privacy issues come to be a challenge as our understanding of genetics advances?

8.6.3 Why might we end up with a world of genetic "haves" and "have nots"? Won't we all benefit from these genetic medical advances? Explain your answer.

In every chapter in your text, the following five sections appear as special boxes within the chapter. You should refer to those as you try to answer the review questions here, which focus on the main point(s) of the section.

8. Biology in Perspective

The DNA of genes directs the production of proteins that have particular functions in a cell. Genes form networks, the proteins work together, and the whole process is sensitive to the environment, including our lifestyle choices. We can manipulate genes in people, collect stem cells from human embryos, and clone. There is still a lot we don't know, and there will be unintended consequences, so we will have to define ethical and moral boundaries to guide this work as we move forward.

Question to consider:

What are some of the things we're able to do with our ever-expanding understanding of genes, and what are the challenges we face as a result?

8. Scientist Spotlight: Nancy Wexler (1945–)

Nancy Wexler graduated with a degree in Social Relations and English, but when her geneticist mother was diagnosed with Huntington's disease (HD) in 1968, she began studying HD in graduate school, and her father founded the Hereditary Disease Foundation.

The family made fighting this disease their priority. Wexler is now a professor at Columbia University and president of the Hereditary Disease Foundation.

Question to consider:

What was the discovery that Nancy Wexler's work with HD patients led to?

8. Technology Connection: Who's the Daddy?

In 1944, the famous silent film star Charlie Chaplin, was sued by an actress over paternity. The star, the actress, and the baby all had their blood types tested, and while it was determined that Charlie Chaplin could not be the father, the judge refused to allow blood type as evidence, and Chaplin was ordered to pay child support. This led to a change in the law regarding blood tests and paternity, but blood types can only rule out paternity, not confirm it. Now, paternity can be determined via DNA testing.

Question to consider:

How is DNA fingerprinting better than blood typing for determining paternity?

8. Life Application: Sex Selection

For various reasons, male children are preferred in many cultures. Now that we can determine the gender of embryos, the sex ratio has shifted dramatically in these cultures. This shift has been caused by selective abortions of females following amniocentesis.

Question to consider:

Technology allows prenatal determination of a baby's gender. What is the downside of this?

8. How do we know: How Human Embryonic Stem Cells Can Be Directed to Form Specialized Cells

The first published report of human embryonic stem cells directed to make specialized human tissues was in 2001. Daniel Kaufman and his colleagues at the University of Wisconsin at Madison got human embryonic stem cells to make precursor cells for blood cell production. These are cells that in normal embryos will make the different types of blood cells. This research was important for two reasons. First, it added to our basic knowledge of how to manipulate human embryonic stem cells. Second, it may allow for the development of a new way to generate blood cells for transfusion and transplantation to treat anemias and blood cancers.

Question to consider:

What was the first type of cell produced from cultured human embryonic stem cells, and how would the technology to develop these particular cells be helpful?

Student Study Guide

Chapter 9: Evolution

How do species arise and adapt?

So far, we've looked at how various biological processes work. All of these have been shaped by evolution, so in this chapter, we look at why these processes work the way they do. Evolution helps us understand this, as well as many facts of nature: why organisms fit in so well with their environments, why so many species exist, and why some species are so similar while others are not. Understanding evolution is vital to understanding nature, biology, and medicine, and in this chapter, we'll look at how evolution works and why it remains somewhat controversial, at least in the United States.

9. Case Study: Lactose Intolerance and the Geographic Variation of Human Traits

9. Case Study Summary

Worldwide, most adult humans cannot digest milk and dairy products without suffering intestinal cramping, bloating, diarrhea, and flatulence. We all start out drinking milk just fine, but about 75% of people lose this ability around age 2. Obviously, milk is good for us, even essential for children, so why does this happen?

The sugar in dairy products is called lactose, and it's broken down by an enzyme called lactase. In humans, a gene shuts down lactase production in most adults, making them lactose intolerant. But in about 25% of adults, there is a genetic mutation, or change in DNA, which allows lactase production to continue. This mutation is common in Northern Europeans and their descendants, and also in several ethnic groups from equatorial Africa. Scientists have determined that this recessive mutation became common among Europeans about 10,000 years ago. Why?

This sort of geographic **variation** in genetic traits is common. It turns out that this is a result of relatively recent human evolution, as our ancestors adapted to local conditions in different parts of the world.

Question for Review

Explain what "lactose-intolerant" means physiologically and genetically.

9.1 Learning Objective: How does your body reflect evolutionary history?

Provide three examples demonstrating human life has evolved.

9.1 Summary

If you want to understand why things in biology are the way they are now, you have to consider their history. You can see three interesting examples of this in humans.

1. Mammal testes hang outside the body instead of being protected inside.

Human (and other mammal) testes are located in the scrotum outside the body. It makes sense that they be next to the penis, but it seems like something so important to reproduction should be safely tucked inside the body instead of dangling outside. Also, weirdly, the vas deferens, or sperm duct, which carries sperm from the testes to the penis doesn't connect directly. Instead, it can be more than 10 inches long, curving up around the pelvis and looping around the ureter before heading back down to the penis (see Fig 9.4). Sperm cells mature in their travels here, but it seems like there could be a more efficient design. And why are the testicles are outside anyway?

If you look at the history of the vertebrates, this starts to make some sense. In the embryos of all vertebrates (animals with backbones), the testes form inside the abdominal cavity, and in most vertebrates, that's where they stay. Mammals, however, unlike, say, fish, maintain a warm body temperature, too warm for sperm cells to develop. So in mammals, the testes move outside the core of the body to keep cooler. And as this happens, the vas deferens has to follow along and reach all the way into the scrotum. There's no physiological reason for this, it's a legacy from our cold-blooded vertebrate ancestors.

2. You can get scurvy.

It's long been known that if you don't get enough vitamin C in your diet, you can get the disease scurvy. This results in spongy, bleeding gums and eventual tooth loss, and open pus-filled wounds. Most mammals, however, can make their own vitamin C using a series of eight enzymes to modify glucose, a simple sugar. Humans and the other primates are missing the last enzyme. We have the gene that makes it, but a mutation has rendered that gene nonfunctional in all humans and primates. A few bats and guinea pigs also can't make their own vitamin C, for a different reason, but otherwise all other mammals can. Primates eat a lot of fruit, so they get plenty of vitamin C, otherwise this mutation would be lethal.

A gene that is mutated to not work is called a pseudogene. Modern molecular techniques have shown scientists that these are quite common, so they can be used to trace ancestry. If one species shares a particular mutated pseudogene with another, it's unlikely that same mutation happened twice. Instead, those two species probably shared a common ancestor with the mutation. So it is with the primates.

3. Your eye is backwards.

All vertebrates have essentially the same eye structure, and they're all organized backwards (see Fig 9.7). Light enters the front of the eye and hits the photoreceptors—rods and cones—in the retina at the back of the eye. These receptors send impulses to nerves that carry them to the brain for interpretation, but the nerves actually lie in front of the retina, potentially blocking light from hitting the rods and cones there to sense the light. And where those nerves exit the eye, they must pass through a hole in the retina, or a "blind spot" where there cannot be photoreceptors. Our eyes work pretty well, but they could work better. In fact, the eyes of squid, octopus, and cuttlefish are much like ours, but their optic nerves are out of the way behind the retina, so they have no blind spot. Why?

Again, it's an accident of history. Eyes developed in small steps over a long period. Each step made the eye a bit more functional, but each step had to build on what was already there. If the nerve cells were out in front of the photoreceptors in one step, they had to stay there in the next step. Our eyes have a history.

Recognizing that the structures in our bodies have a history means recognizing that we've changed through time, or evolved. Evidence of this, and much other evidence, has convinced biologists that evolution is a fact of nature.

9.1 Questions for Review

9.1.1 In the table below, summarize three examples illustrating that humans have evolved:

Example	How are humans different from other mammals in relation to this feature?	How does human evolutionary history explain this difference?
1. Human testes		
2. Vitamin C		
3. Eye organization		

9.1.2 Explain how these examples illustrate that humans have evolved.

9.2 Learning Objective: What convinced Darwin of the fact of evolution?

Detail the discoveries that led Darwin to change his mind about evolution.

9.2 Summary

Charles Darwin, the man most associated with the idea of evolution, initially believed, like most people of his era, that God created all forms of life in much the same form as they appear today. As a naturalist on a voyage on the H.M.S. *Beagle*, however, he studied the plants and animals of South America, and he saw evidence that species change and that new species could arise from existing ones. There were three main lines of evidence:

1. Biogeography

Biogeography is the study of where different species live in the world. For example, there are rainforests on nearly every continent, and if species were created for rainforests, you might expect similar organisms in rainforests worldwide. But Darwin saw that the South American rainforest species had more in common with species from other South American habitats than with species from rainforests on other continents. He also found fossils in South America that were similar to species living there, but not quite the same. Darwin reasoned that one way to explain this was that species on a continent have a history together. They've changed over time to survive as they moved into new habitats or as old habitats changed. He also reasoned that this process must take awhile.

2. Endemic Species (the Galapagos Islands)

During this voyage, Darwin visited the Galapagos Islands, about 600 miles offshore west of Ecuador. On these islands, he found a number of endemic species—species that are found nowhere else. One of these species was the finch, a small bird. Darwin found 14 varieties of finches on the different islands. They varied in size, bill structure, body and tail shape, and behaviors. For example, finches typically have strong beaks for cracking seeds, and some of Darwin's finches did this, but others ate fruit, drilled holes in trees or probed with sticks looking for insects, or even pecked flesh of other animals to drink their blood. But they were all still clearly finches (Figure 9.12). Darwin wrote that "...one species had been taken and modified for different ends."   In other words, the original finch species had changed over time to suit the various habitats on the different islands.

Given this evidence, Darwin changed his mind and concluded that species change through time, sometimes becoming new species. That's how science works. Ideas change according to evidence. Darwin spent 20 years considering this evidence and developing an idea of how the process of change might work. He laid it all out in his book *The Origin of Species by Means of Natural Selection.*

Natural Selection

Natural selection is a simple idea. Individuals with certain traits do better than those with other traits. They are more likely to survive and produce offspring. Through heredity, those offspring will tend to also have those beneficial traits, and that's the way the population will go. Natural selection is based on three observable facts of nature:

1. Individuals in a population vary; they differ from one another.

2. Variations are inherited; parents pass some of their variation on to their offspring.

3. In any population, many more offspring are born than survive to adulthood; there is competition and a struggle to survive.

Point 3 suggests that in most species, most offspring die, so what determines who survives? It could be random, but if the individuals vary (point 1), some will be better suited to survive than others given their particular environment. The survivors reproduce and pass on the very traits that helped them survive (point 2).

As a result, species will change over time. Each generation will differ a bit from the previous one. It will be better adapted to its environment. Different environmental conditions would demand different adaptations such that a species in two environments could adapt independently and split into two species.

Darwin described the change in species over time as *descent with modification*. Natural selection was his way of explaining how the change occurred. At the time most scientists found Darwin's evidence for evolution to be overwhelming, but without an understanding of the genetics of heredity, Darwin couldn't adequately explain species variation, nor how that variation was inherited. And his competitive world portrayed nature as rather heartless, which contradicted the prevailing idea of a harmonious world created by God. It took 50 years for the idea of natural selection to be accepted.

The Modern Synthesis

Do you remember Gregor Mendel and his pea plant experiments from Chapter 4? Once his work was discovered, genetics provided the mechanism by which natural selection could work. Traits are controlled by genes. Genes exist in different forms, or alleles, and those genes get passed from parent to offspring. Organisms can vary and those variations can be inherited. Natural selection explains how allele frequency, the number of particular alleles in a population, changes over time. This joining of genetics with natural selection is called *The Modern Synthesis*.

One result of this synthesis of ideas was that mathematical equations could be developed that would predict how a population would change given a particular set of circumstances. Experiments have shown that natural selection works just as Darwin thought it did. By the 1940s, evolution by natural selection was accepted by scientists as the important and powerful theory it is, supported by much evidence.

9.2 Define these Key Terms

biogeography

endemic

natural selection

variation

Modern Synthesis

Allele frequency

Darwin's finches

9.2 Questions for Review

9.2.1 Describe the observations Darwin made that convinced him that species change through time.

9.2.2. What are the three requirements for natural selection to occur?

9.2.3 Why did scientists not initially accept the idea of natural selection?

9.2.4 How did the work of Mendel contribute to the natural selection becoming an accepted idea?

9.3 Learning Objective: How do humans adapt to their environment?

Describe how humans adapt to their environment using specific examples.

9.3 Summary

So far in this chapter, we've seen that humans vary geographically in their lactose tolerance and also in their tolerance of certain treatments for malaria. In this chapter, we look at why these traits show up where they do.

1. **Lactase Persistence**

Since mutations occur randomly, it's likely that the mutation that allows lactase production to continue in adulthood has shown up many times in human history, but it never spread to a population until about 10,000 years ago in Northern Europe. Why then and there?

For most of human history, we've been hunter-gathers, not farmers. So once infants were weaned, nobody was drinking milk. There weren't any domesticated animals to provide it. Being able to produce lactase and digest lactose as an adult wouldn't have made any difference in one's survival, so the trait wouldn't have "caught on." If milk were available from domestic animals, it would provide nutrition, maybe in tough times. If you could digest it, you wouldn't starve and you could continue to produce offspring; they would inherit the mutation and do better, too. So the trait would spread in the population. Guess where and when we started dairy farming? Northern Europe, 10,000 years ago. Northern Europeans and their descendants in North America and Australia can mostly all digest lactose. Same with a few tribes of equatorial Africa who have a long history of herding dairy cattle.

This type of change resulting from a specific environmental condition is an **adaptation.** The environment didn't cause the mutation, but it turned out that the mutation was helpful in that particular environment.

2. **Malaria and Oxidizing Drugs**

The introductory case study in this chapter describes how some people of African and Mediterranean descent are killed by certain treatments for malaria. Why is this?

We saw in Chapter 4 that the prevalence of malaria in certain regions of Africa has resulted in natural selection for the sickle cell trait, because it confers some resistance to malaria. As it turns out, malaria has caused several other adaptations to evolve. One of these, an enzyme deficiency, reduces the red blood cell's ability to use sugar as an energy source.

This hinders the reproduction of the malarial parasite infecting the red blood cell. We've developed oxidizing drugs that use this same strategy to treat malaria. These drugs generally work well, but if a person with the mutation takes an oxidizing drug, it's too much and may kill the person. As it turns out, our evolutionary history can have serious medical consequences, so it's important to consider it.

9.3 Define this Key Term

adaptation

9.3 Questions for Review

9.3.1 What about human evolutionary history explains why Northern Europeans and their descendants can usually digest lactose?

9.3.2. Would it be accurate to say that farming dairy cattle caused the mutation that allows an adult to digest lactose? Why or why not?

9.3.3. In areas where malaria is common, doctors have to be careful about which drugs they use to treat the disease. Why is this?

9.4 Learning Objective: How does natural selection produce adaptations?

Explain how natural selection produces adaptations, and outline some limits of natural selection.

9.4 Summary

We've seen several examples of how humans have adapted to their environment, but we've not looked at how natural selection results in adaptations. Examples illustrating how this happens are rare because natural selection takes a long time. In this section, you'll learn about one famous study which has been going on for 40 years already and has shown us definitive things about natural selection, but is still a work in progress.

The Grant's Galapagos Finches

Since 1973, scientists Peter and Rosemary Grant have been studying the medium ground finches on the Galapagos island of Daphne Major. Every finch there has been identified over the years with a distinctive leg band. Scientists know every bird: age, weight, beak size, parents, mates, offspring. These birds eat seeds which they crush with their beaks. Beak size is variable, so some birds have slightly wider beaks and therefore can crush bigger, tougher seeds. Daphne Major is a fairly dry island normally, but there is enough rainfall to support a diversity of plants, and a variety of seeds are typically available for the finches to eat.

Between 1976 and 1977, there was a drought on Daphne Major during which only the mostly large, hard seeds were available. More than 80% of the finches died. This set the stage for a test of natural selection. Remember that natural selection requires three things to be true:

1. Individuals in a population vary: the beak depth of the finches varied.

2. The variations are inherited: the Grant's determined that parent's beak size determined the beak size of their offspring.

3: There is competition to survive: with a reduced food supply due to drought, the finches had to compete and many died.

Figure 9.21 shows the original range of beak size. Figure 9.22 shows what happened to the beak size in the population of finches as a result of the drought. They got bigger because in an environment of mostly bigger, harder seeds, only the birds with beaks big enough to eat those seeds could survive. Only those birds produced offspring. Those offspring had bigger beaks like their parents, and so that's the way the population evolved.

Note that the population of finches evolved. An individual can't evolve. Once you have a beak, that's the size it stays. Also, this process can go either direction. Daphne Major has both periods of drought and periods of wetter seasons. The Grants have found that during droughts, population beak size increases, and it decreases during wet seasons when many smaller seeds are available again. At those times, there's no great advantage to having a bigger beak. There is food suitable for every size beak, and a variety of beak-sized birds survive.

Fitness

Natural selection is sometimes described as "survival of the fittest," but scientists have a very specific definition of fitness. It is *the number of genes an organism passes along to the next generation.* This is something that can often be measured, and it doesn't necessarily refer to the biggest or strongest or prettiest being most fit. It's whoever is best adapted to the environmental conditions at the time. The right adaptation increases survivorship, which increases reproduction, which increases the number of genes an organism leaves in the next generation.

Of course, your offspring get half of their genes from you, but you share genes with all your relatives because you have a common ancestor (see Figure 9.23). Scientists' definition of fitness rewards passing on genes, and it doesn't matter who happens to have them, so all your relatives, or kin, "count" in your fitness. Therefore, it's to your evolutionary advantage to help your offspring, of course, but also your siblings, grandchildren, and cousins. We see helpful behavior toward family in humans and other social species. This selection for behavior enhancing the survival and reproduction of relatives is called **kin selection.**

Limits of Natural Selection

It's important to remember that natural selection can't do everything. There are some limits:

1. Natural selection doesn't anticipate the future. Traits are selected that are advantageous given current conditions, not that might be advantageous later, although sometimes that turns out to be the case.

2. Natural selection can only select from variations that are already present in the population. Natural selection didn't create bigger beaks in the finches during the drought. It just selected those finches that already had bigger beaks. Individuals are constrained by their evolutionary ancestry as well. An engineer could have designed better eyes for us, but we don't have an engineer. We just have natural selection working with what it has, selecting the optimal traits of those available.

9.4 Define these Key Terms

fitness

kin selection

9.4 Questions for Review

9.4.1. Explain how the finches on Daphne Major satisfy each of the three requirements for natural selection to occur.

9.4.2 Review Figure 9.21, which shows finch beak size before and after the drought. Explain what happened and how this illustrates adaptation by natural selection.

9.4.3. Explain why beak size in the finches decreases during wet seasons following dry seasons.

9.4.4 What is meant by "fitness" in evolutionary terms? How is one individual more "fit" than another?

9.4.5. If natural selection "chooses" the best traits for current conditions, why aren't organisms perfectly designed?

9.5 Learning Objective: What Are Random Events in Evolution?

Identify the major causes of genetic change in populations.

9.5 Summary

Natural selection allows organisms to adapt to certain environmental conditions, but random events can also cause the traits of a population to change (and remember that any kind of change in traits or genes is evolution). There are two main random events that can drive evolution.

1. Mutations

Remember that a mutation is a random mistake made in copying DNA. These mistakes are rare, but organisms have so much DNA that any one individual is likely to have a couple mutations. These provide the variation on which natural selection can work. There are three possible outcomes:1. If the mutation is beneficial, that trait will be selected for, the individual will survive and reproduce successfully, and the mutation and trait will pass into the population (like the ability to digest lactose). 2. A harmful mutation will lower the fitness of the individual, so it will not get passed into the population. 3. Most mutations are neutral, and natural selection doesn't affect them one way or the other.

2. Genetic Drift

Mutations occur at random, so a population sometimes doesn't change much at all for a long period, and sometimes there are a lot of mutations over a short period and the population changes. This random change is called **genetic drift** because the genes drift randomly one way and then another, not due to selection.

The **founder's effect** is a special case of genetic drift. In this case a small subset of a large population is isolated from the main population. For example, a particular species of butterfly comes in grey and blue colors. A storm blows a subset of this population off onto an island. By chance, most of the butterflies that end up on the island are blue and they now can't interbreed with rest of the population. Right away, then, the island population is different because the "founding" members are different. They're mostly blue rather than being gray or blue. And from then on, they'll adapt differently to the different island environment.

The Founder's Effect in Humans

There are many examples of the founder's effect in human evolution. Our earliest ancestors dispersed from Africa, and individual small groups headed off to new places, apart from the main groups. The native residents of Asia, Australia, Europe, and the Americas all descended from small groups of people that started off genetically different from the main group and then adapted to their new, unique environments. This is why the native people of these regions differ in skin color, height, and susceptibility to certain diseases like sickle cell (remember that?).

Another example comes from the Amish population in Eastern Pennsylvania. This group descends from perhaps 200 German immigrants dating back to the 1700s. One couple in this original group carried a mutation for Ellis-van Creveld syndrome, which has several symptoms, such as having more than five fingers or toes and signs of dwarfism. It's noticeable but not fatal. The trait was passed to their offspring, and because the population was small and isolated. As other Amish married into this family, the trait spread. Today, 5 of every 1000 infants of the Old Order Amish in Pennsylvania have the syndrome, compared to 1 in 60,000 in the general population. It's obviously not an adaptation. A couple of the settlers just happened to bring it with them.

9.5 Define these Key Terms

mutations

genetic drift

founder's effect

9.5 Questions for Review

9.5.1 How do mutations drive the process of evolution?

9.5.2 What happens to a beneficial mutation in a population? A neutral mutation?

9.5.3. Explain how random mutations leads to genetic drift. How is that evolution?

9.5.4. Using an example, explain how the founder's effect results in evolution in a population.

9.6 Learning Objective: What is the evidence for speciation?

Summarize the evidence for speciation.

9.6 Summary

So far in this chapter, we've seen lots of evidence that species change over time, but different is not the same thing as new. How do scientists know that these processes can eventually result in whole new species, or **speciation**?

What is a Species?

Scientists consider two organisms to be of the same species if they can interbreed. For this to work, they have to share the same set of genes. Note that this is not the same *alleles*. There is a lot of variation within those genes, but the genes themselves mostly match up, or else the organisms couldn't produce offspring. Individuals of different species cannot produce offspring together. They are genetically isolated from one another, which means each follows its own independent evolutionary path. This is the **biological species concept**.

This definition of species mostly works, but not always. Lions and tigers can interbreed if you put them together in an artificial situation. So can dogs and wolves. There are plenty of hybrid plants. Bacteria swap DNA asexually, so they don't really "breed" at all. And how does a scientist decide if a fossil is a new species or the same one he or she already has? You can't get fossils to breed either.

This rather messy situation actually supports the idea that species form by branching off from one another, but it takes a long time, so we find these "in between" situations. Some are just starting to split. Dogs did not split off from wolves all that long ago, so we might expect that they would not be completely genetically isolated yet, and eventually they will not be able to interbreed. The clearly defined species have been separate for a long time. If every species was created completely separately from every other one, it would probably be really easy to distinguish them.

Evidence for Speciation

DNA is passed from one generation to the next, but it accumulates mutations. If you compare one organism's DNA sequence (the order of the nucleotide bases) to another's and they are similar, you can conclude that the two organisms are closely related. If the sequences are quite different, the two organisms haven't shared genes in a long time, so they are more distantly related. Scientists use this information to construct family trees for whole groups of organisms. You can see an example of one of these evolutionary trees in Figure 9.31.

This tree is for shrimp species from both the Pacific and the Caribbean side of Panama. Notice that the species sort into pairs with one member from the P side and one from the C side. In each group, P and C are closely related, even though they're from different oceans on different sides of Panama. Oddly, these pairs are more closely related to each other than to other species from the same ocean which live just a bit north or south. How can this be?

It turns out that until the Isthmus of Panama rose up out of the ocean about 3 million years ago, North and South America were not connected and the Caribbean and Pacific flowed together. At that time, there were probably seven species of shrimp. When the Isthmus separated the Pacific from the Caribbean, it also split the shrimp populations, and the shrimp on the Pacific side evolved a bit differently from the Caribbean side, genetically diverging as they adapted to different conditions on either side of the Isthmus. By now, they've diverged into separate species, but their DNA still shows that they were once related.

9.6 Define these Key Terms

speciation

biological species concept

evolutionary trees

9.6 Questions for Review

9.6.1 How do scientists tell how to put together an evolutionary tree? What do they look at which indicates how closely related one species is to another?

9.6.2 How do scientists know that the Pacific shrimp and the Caribbean shrimp, which are isolated from one another on either side of the Isthmus of Panama, are closely related? What did they look at and what did they find?

9.6.3 Explain how the case of the Caribbean and Pacific shrimp illustrates the process of speciation. What caused the original seven shrimp species to more than double?

9.7 Learning Objective: How do new species arise?

Outline the three stages by which new species arise.

9.7 Summary

Darwin recognized that two populations within one species might live in different areas and evolve independently, each adapting to their own environment. These two populations could become so different over time that they would be different species. It turns out that this process of speciation happens in three stages.

1. Genetic Isolation

This is illustrated in Figure 9.33. Something divides populations such that they can no longer interbreed. Often this is a geologic feature like the Isthmus of Panama, the Grand Canyon, or a river. Or a small number of individuals can migrate to a new area, separate from the original population. The populations are physically separated, and since they can no longer interbreed, they are genetically isolated.

2. Genetic Divergence

Initially, the founder's effect will result in the new population and the original being genetically different from one another. Mutations will naturally occur over time, differently in each population. Natural selection will increase or decrease the frequency of some of these mutations differently in the two populations, as the environmental conditions are likely to be different. Over time, the populations diverge genetically, and probably also structurally and behaviorally.

3. Secondary Contact

In stage 3 of speciation, the separated populations come back into contact, providing an opportunity for interbreeding. There are three possible outcomes at this point (illustrated in Figure 9.35).

1. Despite their genetic divergence, the two populations can interbreed and probably will. No speciation has occurred.

2. Genetic divergence is so great that the populations don't even try to mate; therefore, speciation has occurred and there are now two species where there was one.

3. It's also possible that the populations will interbreed and produce hybrid offspring. These hybrids are likely to be less well adapted to either of its parents' habitats, or they may be less viable or less fertile or sterile. Because hybrids are less fit, natural selection will work against parents that produce hybrids. This type of natural selection is called reinforcement because it reinforces the separation of the two populations. Reinforcement might select each population to develop different mating rituals, to mate at different times or places, or to develop some traits which ensure prospective parents don't mistakenly interbreed with the "wrong" population. Interbreeding, therefore, no longer occurs, and according to the biological species concept, we have two species.

9.7 Define these Key Terms

genetic isolation

genetic divergence

reinforcement

9.7 Questions for Review

9.7.1 Draw a diagram or flow chart showing the steps leading to speciation.

9.7.2. Why is it necessary for two divided populations to come back into contact in order to determine whether they have diverged into two different species?

9.7.3 Natural selection can reinforce the separation of two populations even if they can interbreed to produce hybrid offspring. Explain how reinforcement works.

9.8 Learning Objective: Why is it so difficult for the public to accept evolution?

9.8 Summary

When you look at the section of the chapter *Our Life Application: Public Acceptance of Evolution*, you'll see that in most countries the public has no real problem accepting evolution. In the United States, however, only about 40% of the population accepts it, 40% doesn't accept it, and 20% don't have an opinion.

Scientists are nearly unanimous in their acceptance of evolution as a fact of life, as well supported by evidence as the fact that the Earth revolves around the sun. A number of religions have formally adopted the position that evolution is completely compatible with their spiritual teachings, but certainly not all have done so, and the arguments against evolution are usually religious in nature.

It's a bad idea to use evolution to support social and political ideology because it can be used to justify aggression, exploitive capitalism, and genocide, all in the name of "survival of the fittest." Scientists would caution against applying the principles of evolution to our society, because human society is too complex to be explained by biology alone.

The fact remains that evolution explains what we see in the world around us: why some species are more closely related than others, and why they're adapted to their habitat. We now have more than a century of data from every scientific discipline which points to the fact of evolution. To argue that evolution is wrong is to argue that science is wrong, evidence doesn't matter, and personal belief carries more weight than evidence.

9.8 Questions for Review

9.8.1 Why do the vast majority of scientists accept evolution as fact?

9.8.2 Why doesn't it make sense to apply evolutionary principles to human society?

9. Biology in Perspective

Summary

Darwin's work and the work of countless scientists since has confirmed that species adapt by natural selection. Species arise in a three-step process that includes both natural selection and elements of chance. This framework explains two of the most prominent observable facts of nature: the diversity of species and the fit of species to their environment. As a result of evolution, living things have a history.

Despite the evidence, evolution continues to drive a long-standing conflict between science and religion in the United States. Education about both religion and evolution is the best way to reconcile belief with science.

Questions to consider

Summarize how natural selection works.

How does natural selection explain the variation we see in humans from one region to another?

9. Scientist Spotlight: Sir Ronald Aylmer Fisher (1890–1962)

Summary

R.A. Fisher was one of the main scientists behind the development of The Modern Synthesis. He worked as a statistician at Rothamsted Agricultural Experiment Station, where he developed statistical techniques useful in experimental crop breeding. In 1930 he was appointed Professor of Eugenics at University College and also published an influential book called The Genetical Theory of Natural Selection. He held the extreme view that the way to improve humans was to prevent biologically defective people from having children. His teaching and writing were difficult for most people to understand, but his contributions to statistics and biology were significant, far reaching, and long lasting. He has been called "Darwin's greatest 20th-century successor," and "a genius who almost single-handedly created the foundations of modern statistical science."

Questions to consider

What is R.A. Fisher known for?

Aside from the ethics of the issue, why does it make little sense that only having wealthy and educated people have children would increase the genetic quality of the human population?

How Do We Know?: Constructing Evolutionary Trees

How do scientists study the lineages of species?

Summary

One of the most important consequences of speciation is that all species share a common ancestor. Evolutionary biologists construct evolutionary trees to study lineages of species just as genealogists construct family trees. Because relatedness is determined by how similar DNA is, one of the best ways to group species is by DNA similarity. The longer the time since the common ancestor, the more mutations there are. The more mutations species share, the more closely related they are, and the closer they are on the tree. The process of then grouping species according to relatedness and building the tree is highly mathematical, and these days the calculations are done mainly by computer.

Technology Connection: GenBank

How are online tools being used to further scientists' analysis of genetic information?

Summary

Scientists can now extract DNA from the cells of the organisms they study and rapidly determine the sequence of nucleotides or bases in each segment of DNA. Anyone can upload this information to a huge, and constantly expanding, database called GenBank. GenBank also stores protein sequence data. This database is maintained by the National Center for Biotechnology Information (NCBI) so that everyone can access it.

Questions to consider

If you were interested in the family tree or evolution of dogs relative to other mammals, explain how you could use GenBank and the NCBI tools to investigate this.

Scientists often want to keep their own data to themselves. Why do you think they're happy to share DNA sequences with everyone in a nationally maintained databank?

Life Application: Public Acceptance of Evolution

Why does the United States have a significantly lower public acceptance of evolution

than its European allies or Japan?

Summary

Surveys consistently show that fewer people in the United States accept the idea of evolution than in most other countries. A sample of these survey results is shown in Figure I A. In 2012, a Gallup poll revealed that only 15% of Americans believe in atheistic evolution, 32% believe God guides evolution, and 46% believe in creationism rather than evolution at all. One reason for this is that the United States has a stronger presence of Christian sects that believe in biblical literalism. A second reason is that evolution has become politicized in the United States. The conflict is ongoing. Probably, as people learn more about science in general, and evolution in particular, there will be fewer misconceptions about both, and fewer disagreements.

Question to consider

Evolution is generally not well accepted in the United States compared to other countries. What are two contributing reasons for this?

Student Study Guide

Chapter 10. The Evolution of Disease

Why Do We Get Sick?

The answer is complicated and interesting. Pathogens, which are disease-causing bacteria, fungi, or worms, invade our bodies and do damage, as do viruses. But thousands of bacteria live on us all the time and help keep us healthy. Pathogens evolve in response to our defenses, but understanding that process may lead to better preventatives, treatments, and cures.

10. Case Study: Deadly Malaria

Why is Plasmodium, *the pathogen that causes malaria, not as easy to kill as it was in the past?*

Case Study Summary

Malaria has probably killed more humans than any other disease. About 300 million suffer from it annually, and about 1.5 million die.

There are four species of *Plasmodium* that infect humans, but *P. falciparum* underwent a population explosion about 7,000–10,000 years ago and is now the most deadly, accounting for more than half the cases of malaria and 99% of the deaths worldwide.

One reason *Plasmodium* is so deadly is that it's not as easy to kill now as it used to be. We used to use quinine, and now we have a dozen different drugs to treat malaria, but none is effective against every type. Different types of malaria are caused by different strains of *Plasmodium,* each with different genetic variations, some of which make them resistant to multiple drugs.

We, too, have been evolving to fight back. The sickle cell mutation makes blood cells less vulnerable to malaria. There are four different mutations that cause red blood cells to become sickle-shaped, and there are about a dozen other mutations that make red blood cells partially resistant to *Plasmodium.*

Evolution of resistance by the *Plasmodium* has made malaria deadly, but so have changes in human behavior that allow it to spread.

Questions for Review

Why is malaria difficult to treat?

How are humans evolving in response to malaria?

10.1 In What Ways is Your Body an Ecosystem?

Learning Objective 10.1:

> *Recognize and provide examples of the different types of species interactions*

10.1   Summary

Only about 10 trillion of "your" cells are genetically you. The other 90 trillion belong to bacteria, fungi, yeast, mold, and multicellular organisms like mites and worms. Most of these live on our skin, in our mouth and digestive tracts, urinary tracts, and nose and lungs.

Ecology is the study of interactions among species and their physical environments. Each of us has our own particular set of species and these form ecological communities throughout our bodies. Your dry forearm has "desert" adapted species. The more "tropical" communities occupy your warm, moist underarms and groin. More than 700 species live in your mouth and more than 800 in your colon.

Individuals in these communities survive, reproduce, compete, so they evolve and adapt over time. Your body is an ecosystem.

Most of our resident species appear to have no effect on us at all. Eyebrow mites eat dead skin cells and excess oil, and we don't know they're there. This type of interaction where one species benefits and the other is neither helped nor harmed is called commensalism.

Our resident species compete for food and space. Some secrete toxic compounds that kill other species. Some use up critical nutrients or make the habitat too acidic for competitors. Some in the intestine form a slippery film so other species can't attach.

Many species that live with us are mutualistic, meaning we both benefit from the interaction. Species in the digestive tract help digest our food. They can also outcompete potential invaders like pathogens.

Many of these species cause us no trouble most of the time, but if conditions are right and their populations get too large, they can cause problems. For example, *Pseudomonas aeruginosa* can live almost anywhere, including on you and cause no problem. But if a person is already sick or has a weak immune system, the population can grow to a particular level, or quorum, and it can overwhelm the body. *P. aeruginosa* can cause pneumonia, urinary tract infections, septic shock, tissue necrosis (where the tissue dies) or death.

In some cases, a species always causes disease. A virus, for example, generally can't reproduce without killing the cell it infects. And many pathogens are truly invasive. They don't normally live on us but we inhale or swallow them, or they get in through wounds or by burrowing, and cause disease.

10.1 Define these Key Terms

pathogen

ecology

commensalism

mutualism

10.1 Questions for Review

10.1.1 How do we and our resident organisms comprise an ecosystem?

10.1.2 Using our resident organisms, explain an example of commensalism.

10.1.3 Using our resident organisms, explain an example of mutualism.

10.1.4 What is meant by a quorum in a bacterial community?

10.1.5 What are some of the ways that our resident community of organisms keeps unwanted organisms from settling in?

10.2 Why Do Diseases Evolve Resistance to Antibiotics?

Learning Objective 10.2:

Describe the process by which some disease organisms increase their resistance to antibiotics.

10.2 Summary

In 1928, a Scottish scientist, Andrew Fleming, discovered that one of his bacterial cultures had been contaminated by a mold, *Penicillium notatum*. He noted that bacteria did not grow right near the mold. He showed that the mold produced a substance he called penicillin which killed bacteria. A drug that kills bacterial pathogens is called an antibiotic, and we've since developed many others. The problem is that after a few years of use, strains of disease evolve that are resistant to new antibiotics.

What happens is that the antibiotic kills the majority of the pathogens in a population, but a few are already resistant, and those survive and reproduce. Their offspring are also resistant, and after short time, all the cells remaining are resistant. It's natural selection at work. This happens with insects and insecticides, too.

For any organism to produce an antibiotic, like *Penicillium* does, it must be resistant to it, or it would kill itself while it was trying to kill its competitors. There must be resistance genes that mutate to produce proteins that make a cell resistant. There are four ways this can work:

1. They change the cell membrane so the antibiotic can't get in.
2. They modify an enzyme so it breaks down the antibiotic.
3. They alter the structure of the antibiotic's target (what it attacks in the cell) so it's no longer vulnerable.
4. They form a pump that pumps out the antibiotic before it causes damage.

These mutations would be rare, but bacteria have been around a very long time, and they exist in very large populations, so there are a few of these mutations in any bacterial population. In addition, bacteria can trade genes through horizontal gene transfer, so a resistant bacterium can share its resistance with its neighbors. Horizontal gene transfer happens in three ways:

1. Transformation: DNA released into the environment by bacteria is absorbed by other bacteria.
2. Transduction: DNA is transferred from one bacterium to another by a virus.
3. Conjugation: Two bacterial cells join together and transfer DNA from one to the other.

In all of these cases, the descendants of the altered bacteria are resistant.

Because it takes energy to resist antibiotics, cells that are resistant have less energy for other things, like reproduction and growth. So if there is no antibiotic present in the environment, the non-resistant cells have the advantage, and they end up being the majority of the population. This is why it is important to take antibiotics exactly as long as prescribed. The dosage is long enough to kill the vulnerable cells but short enough to prevent a population of resistant cells from forming.

Because resistant strains are most likely to develop where there is prolonged use of antibiotics, hospitals tend to harbor many such strains. In addition, there are plenty of weakened patients, susceptible to infection.

Antibiotics are also used extensively in agriculture, sometimes to treat infections or keep pathogens from spoiling crops, but farmers also give low doses of antibiotics to livestock and poultry in order to kill some of the bacteria in their guts so that more food goes to the animals, not the bacteria, and the animals grow faster.

The more prevalent antibiotics are in our environment, the more resistant strains of pathogens develop. It's feared that we will eventually be unable to treat infections with our known antibiotics.

10.2 Define these Key Terms

penicillin

resistance genes

horizontal gene transfer

transformation

10.2 Questions for Review

10.2.1 Explain how penicillin was discovered.

10.2.2 Draw a diagram that explains the evolution of a resistant population of bacteria in the presence of an antibiotic.

10.2.3 Explain the four ways a bacterial cell gene mutation can cause antibiotic resistance.

10.2.4. Describe the three ways a bacterial cell can "share" its antibiotic resistance with others in its population.

10.2.5 Explain why all populations of bacteria are not resistant to antibiotics.

10.2.6 Why are hospitals a good source of antibiotic-resistant infections?

10.3 Why are Some Diseases More Deadly Than Others?

Learning Objective 10.3:

Use the "trade off" hypothesis to explain and predict the spread and impact of a disease

10.3 Summary

Virulence is a measure of how deadly a particular pathogen is. Malaria is highly virulent. The common cold is not. These pathogens are both adapted to best suit their way of life, though. And if a pathogen requires a host for energy and nutrients, it doesn't seem like it ought to kill the host.

Initially, it was thought that the deadly pathogens were new and hadn't yet had time to evolve into milder forms. We've seen this in the myxoma virus, an introduced disease, lethal to rabbits, that was introduced to Australia to combat an out-of-control rabbit population.

The Trade-Off Hypothesis

So it's true that in some cases pathogens are selected for traits that cause less harm to their hosts, but not always. Malaria has been around for 200,000 years and remains deadly. There are two ideas on this:

1. Natural selection doesn't select for "niceness" in pathogens, or anything else. It's all about the traits that allow for maximum reproduction.
2. Pathogens have to both compete and reproduce within a host, but they also have to get to the next host.

In this second case, there is a trade-off between reproduction within the host and transmission to the next host. If a milder disease increases transmission, it will be selected for (like in the case of myxoma and the rabbits). But if killing the host helps the pathogen get to the next host, that's what it does. Pathogens have evolved both ways.

A Disease in Your Body

The original infection may be just a few cells, but the invaders reproduce rapidly. This population growth is what causes harm to the host (you). These organisms sap your energy and do damage to cells and organs. Your immune system has to respond vigorously and sometimes overreacts, which also sometimes increases the severity of the disease.

As these pathogen populations grow, they accumulate mutations and evolve. More efficient ones reproduce more quickly and cause more damage. In addition though, the variation with the largest population is most likely to be passed along to a new host. So fastest growing and most virulent should win. But in some cases, the host needs to survive awhile in order to pass on the infection. So here is the trade-off: grow a bit slower and allow your host to live longer, and your odds of getting to a new host may increase.

Malaria provides a test case for the trade-off hypothesis. It has evolved to be deadly, but it can be transmitted via mosquitoes, so it doesn't matter if its host is almost immediately confined to bed. The mosquitoes can still get around.

This might explain why *Plasmodium* became more deadly about 10,000 years ago. Prior to that, humans lived in small, isolated bands of hunter gatherers. If malaria killed them quickly, it would quickly run out of hosts. But when we started gathering in larger communities to farm, the *Plasmodium* that could reproduce the fastest, and therefore was the most deadly, would have an advantage.

Therefore, the trade-off hypothesis says that diseases that spread most easily will be most deadly. Those carried by insects or water fall into this category. Malaria is one. Bubonic plague, which killed about one third of the human population in the Middle Ages, was spread by fleas. Cholera is spread by water.

Milder diseases like mumps, measles, or chicken pox, have to be spread directly, person to person, by airborne particles or fluids. The people need to be up walking around for this to happen most efficiently.

Of course, this trade-off doesn't always explain things. Tuberculosis and influenza can both be quite deadly and are both transmitted person to person.

Sexually Transmitted Diseases

For a sexually transmitted disease to be transmitted, sex has to occur, so the host has to stay healthy enough for that. The trade-off hypothesis predicts that a sexually transmitted disease will start slow, stay in the host a long time, and at least start out being fairly mild. Two examples show this pattern:

1. Chlamydia is the most common STD in the United States, which means it has been successful from an evolutionary standpoint. It's hard to get a count, however, because about half the men and three quarters of the women with the disease don't know they have it. It takes at least 3 weeks for any symptoms to show up.
2. Syphilis is another STD that may not show symptoms early on. Sores or rashes may occur in the first 2 months following infection, but they might not show up at all, or they might not be noticeable. People don't get treatment and pass the disease on. AIDS follows this same pattern.

10.3 Define these Key Terms

transmission

virulence

10.3 Questions for Review

10.3.1 What two "goals" of evolution are in conflict in pathogens according to the trade-off hypothesis?

10.3.2 Given #1, describe how the trade off hypothesis explains why some diseases are more virulent than others. Describe how this works in at least one example of a very virulent disease and a very mild disease.

10.3.3. Explain from an evolutionary standpoint why STDs are very often "silent." What's in it for the pathogen?

10.4 Where Do New Diseases Come From?

Learning Objective 10.4:

Consider the process of mutation and natural selection and predict likely sources of new disease.

10.4 Summary

New diseases are all discovered routinely, and there's always a concern that some new disease will become the next global epidemic.

Sources of New Disease

There are three sources of new disease:

1. Species that normally live in your body can evolve a new strain that causes disease.
2. You can catch the disease of another species.
3. Species that normally live in soil or water can invade your body.

Most new diseases come from pathogens that jump from other species, mostly mammals and birds, to us. Every species has its own set of species living with it, and ones jumping from one species to another is not much different than a species migrating to a new region. If it can adapt to its new home, disease-causing or not, it can thrive. This adaptation could be a mutation of the disease organism, or it could be a change in human behavior that allows it to make the jump.

Stages of a New Disease

Scientists who study evolutionary medicine have identified four stages that a new infectious disease typically moves through when it moves to humans:

1. Exposure: Something has to happen that brings the pathogen into contact with humans. We're exposed to pathogens all the time, and our immune system fights off most of them, so only a few make it to stage 2.
2. Infection: This occurs when the pathogen is able to invade our body and reproduce inside.
3. Transmission to Others: The pathogen has to be able to reproduce many offspring, and those must get to appropriate locations and tissues to get out of the body. The lungs or digestive tract would work, for example. And the pathogen has to be able to evade our defenses long enough for this to occur.

4. Epidemics: This is the final stage at which the disease can spread widely, and a greater number of people over a larger area are infected than would be expected. Our pathogen has found the best balance between reproducing and spreading to new hosts.

A disease can get to epidemic stage through natural selection by selecting the most successful strains of the pathogen until the population is suited to infect and transmit. Or it can reach the epidemic stage because of a change in human behavior or society, like living more closely together in cities, or having multiple sexual partners.

HIV/AIDS

AIDS, Acquired Immune Deficiency Syndrome, is caused by the Human Immunodeficiency Virus, HIV. It's been in the human population for about a century, although the first reported cases in the United States arose in 1983. HIV attacks and kills the blood cells that help make antibodies, compounds that kill pathogens. This makes the infected person more vulnerable to diseases that would not kill a person with a healthy immune response.

HIV-1 jumped from chimpanzees to humans and is the more virulent form. HIV-2 victims live longer and show symptoms later, but the disease is still fatal. The disease has become a worldwide epidemic.

10.4 Define these Key Terms

infection

transmission

epidemic

antibodies

10.4 Questions for Review

10.4.1 What are the three sources for new diseases?

10.4.2 Draw a flow chart tracing the path a new disease follows to get to the epidemic stage. Describe what is going on at each stage.

10.4.3 Using your flow chart, describe the path that HIV/AIDS took to reach epidemic status.

10.4.4. What does HIV do to a person to make them sick?

10.5 How Can Evolution Help Us Control Disease?

Learning Objective 10.5:

Propose a strategy to reduce the virulence of a disease using evolutionary medicine

10.5 Summary

Antibiotic resistance shows that we can influence the evolution of diseases, so it may be possible to influence it in a positive way. Scientists increasingly believe that integrating evolutionary and medical research will bring about benefits.

Antibiotic Resistance

One thing we could do is reduce our use of antibiotics: overuse leads to more resistant pathogens. We could target them more narrowly toward the most severe diseases and limit their use in non-health-related activities like agriculture. This will be difficult because pharmaceutical companies make a lot of money from antibiotics, doctors are under pressure to prescribe them, and their use in agriculture to make animals grow more quickly and protect crops may make it difficult for farmers to give them up. There's no way around this process of natural selection, resulting in more resistant pathogens. Antibiotics alone cannot be a long-term strategy for controlling disease.

Vaccinations

Vaccination, which provokes an immune response to a specific pathogen, has been successful in controlling many childhood diseases, and small pox was actually driven to extinction by a global vaccination program. The problem is that some pathogens, like the influenza virus and HIV, mutate so rapidly that the immune system doesn't recognize the new strains if it has been vaccinated against the old strain. That's why you need a new flu shot every year.

In cases like diphtheria, the vaccine targets the disease-causing behavior of the pathogen, not the organism itself.   Diphtheria bacteria produce a toxin that kills cells in our upper respiratory tracts. The diphtheria vaccine stimulates the immune system to make antibodies to the toxin, not the bacterial cell. The cells live, but they're harmless. And because it takes energy to make the toxin, once the toxin is ineffective, the diphtheria bacteria that don't make the toxin grow and reproduce more quickly and outcompete the cells that do.

Controlling Disease by Selecting for a Milder Form

According to the trade-off hypothesis, the most lethal diseases are those that are most easily transmitted. Reducing transmission, then, protects people from being infected in the short term, but in the long term it should select for milder forms of the disease because only those will be transmitted at all.

HIV is spread by unprotected sex and sharing needles. This ease of transmission allows particularly virulent strains of HIV to persist. If behavioral and social practices make it more difficult for HIV to spread, they would make it more difficult for very virulent HIV to persist, and HIV should evolve to a somewhat milder form that might be easier to survive.

10.5 Questions for Review

10.5.1 Explain why antibiotics can't really be a long-term strategy to fight diseases.

10.5.2 What could we, as a society, do that would help reduce the speed of development of antibiotic resistance?

10.5.3 Why do you have to get a flu shot every year, but not a mumps shot?

10.5.4 Most vaccines trigger an immune response against a particular pathogen so that the pathogen is killed off. What's different about the diphtheria vaccine? How does that one lead to diphtheria cells that don't harm you?

10.5.5 Draw a flow chart illustrating the steps that might lead to the development of a milder form of HIV over time.

10. Biology in Perspective

A human is an entire ecosystem of our own cells, bacteria, fungi, mites, and worms. Mostly, we get along fine, but sometimes a pathogen overtakes us, or our own microbes malfunction and become pathogenic. Pathogens can evolve to overcome our defenses against them. New diseases are always emerging. A better understanding of the evolutionary dynamics of disease may provide new approaches to public health.

Question to consider:

Why does it matter that we understand how natural selection drives evolution of pathogens? Explain with an example or two.

10. Scientist Spotlight: Paul W. Ewald

How did a case of diarrhea lead Paul Ewald to a breakthrough in evolutionary medicine?

As a result of a traumatic experience with a pathogen, Dr. Paul W. Ewald's research focuses on host–pathogen interactions, looking at diseases not just from the standpoint of the host but also from that of the pathogen. How did they evolve to be the way they are?

Ewald is now a Professor of Biology and Director of the Program on Disease Evolution at the University of Louisville and has become one of the pioneers of evolutionary medicine. His research focuses on questions of whether some chronic diseases like mental illness, cancer, Alzheimer's, and heart disease are caused by long-term, low-level infections, and whether pathogens can be made less virulent by altering their behaviors or transmission patterns.

Question to consider:

What was it Dr. Ewald wondered, in the midst of an attack of diarrhea, that led him to his current field of research? What kind of questions does he try to answer?

10. Technology Connection: How Vaccines Are Made

How is it possible for scientists to design safe "live" vaccines?

Your immune system recognizes foreign cells and viruses that invade your body. Foreign proteins on their surface indicate foreign DNA, which indicates an external invader. Once that foreign protein is detected, your immune cells make antibodies which are proteins that bind to foreign proteins and "tag" them so other immune system cells will find and ingest them. Your immune system then "remembers" that foreign protein, and having made the antibody once, can quickly make it again if you are infected with the same disease. This is why immunity can last a long time.

Vaccines are composed of the surface proteins from a particular pathogen, fragments of the pathogen's cells, or whole dead or living cells. When these are injected, your immune system recognizes them as foreign, makes antibodies, and you are immune from infection later on. There would be a stronger immune response to the live pathogen, but you'd get sick. However, if the pathogen is grown for many generations in some non-human host (e.g., a chicken egg, another animal, human cells in a lab), natural selection will adapt to these conditions and become less adapted to humans to the point that they don't make us sick and can be used in vaccines.

Question to consider:

How is it possible to make a live vaccine safe to use, and why is it worth doing so rather than using a "dead" vaccine?

10. Life Application: Malaria and DDT

How can we balance savings millions of children's lives with the risks of birth defects, cancer, and damage to the environment that the use of DDT may cause?

DDT is a powerful insecticide that was used during World War II to control mosquitoes, lice, and other insects. DDT, along with screened doors and windows, draining swamps, and better medical care helped eliminate malaria in many developed countries. After the war, it was widely sprayed on crops to reduce insect damage. As it turned out, DDT affected the mammals and birds that ate the insects, and some studies found links between DDT and human birth defects and cancer. In 1962, Rachel Carson published the book *Silent Spring*, which chronicled these effects and led to the modern environmental movement. Ten years later, the United States banned DDT. Since then, we have also learned that insects develop resistance to DDT through natural selection and that DDT persists in the environment long after it loses its effectiveness.

Question to consider:

What is DDT, why is it used, and what are the problems with using it?

10. How Do We Know?: The Many Species That Live on You

How are scientists learning more about the organisms that live on us without growing them in a lab?

Most of the species of bacteria that live on your body are highly adapted to the conditions there and don't grow well, or at all, in a Petri dish in a lab, so scientists couldn't grow them and didn't know about them. Currently, scientists can collect a swab from your skin, your feces, or a cough, and break up the cells they catch. This releases the ribosomal RNA, and because every species has unique rRNA, scientists can determine the species in the sample without having to grow them. We've learned there are thousands of species living on and in us.

Question to consider:

Why is it only recently that scientists have discovered that we have thousands of bacterial cells living on us?

Student Study Guide

Chapter 11 Homeostasis

Why is it important that the body maintain its internal balance?

Our bodies work best when internal conditions are kept at their normal levels. Temperature, water and salt balance, and the amount of dissolved chemicals in the blood are all maintained within a very narrow range, mostly by the brain. The chemical reactions in our bodies work most efficiently under these conditions, and when the balance shifts, things can go very wrong.

11. Case Study: Max Gilpin

Why did Max Gilpin collapse on the football field?

11. Case Study Summary

Max Gilpin, a high school sophomore on the Pleasant Ridge Park football team in Louisville, KY, took to the practice field on a hot, humid day in August 2008. Coach Stinson told the players they'd run sprints to develop stamina and they'd run until someone quit the team. Someone did quit, but not before Max Gilpin collapsed on the field.

Gilpin was a big offensive lineman. Making a 200-lb body run sprints is a lot of work, and muscles doing that work generate a lot of heat. Sweat is the main way the body gets rid of excess heat, but that also involves water loss. When it's hot and humid, it's even harder for the body to lose heat, and with exertion, the body can overheat and dehydrate. Gilpin collapsed on the field and at the hospital his temperature registered 107°. After 3 days, he died from heat-related organ failure: heat stroke.

Coach Stinson was charged with reckless homicide. Players testified that the sprints were unusually severe and that some players were denied water. But an autopsy that might have shown Gilpin was dehydrated was not done. The defense argued that his death was caused by the drug Adderall, whichhe was taking for attention deficit hyperactivity disorder. Coach Stinson was acquitted of all charges.

Question for Review

What is the most likely physiological explanation for what caused Max Gilpin to collapse and die?

11.1 What Is Homeostasis?

11.1 Learning Objective:

Explain why organisms maintain homeostasis.

11.1 Summary

Human body temperature averages 98.6°F or 37°C (Celsius scale) and when we're in good health, it varies by less than one degree. This ability to maintain a constant internal condition in the face of changing external conditions is called homeostasis. There are several other examples in our body besides temperature:

- o   The amount of water in the blood
- o   The concentration of salts and waste products in the blood
- o   The concentration of oxygen in the blood
- o   The concentration of glucose in the blood
- o   The resting rate of the heart and breathing
- o   Blood pressure

In each of these cases, if something happens that changes one of the values, the body has mechanisms to bring it back to within range.

Your metabolism works fastest and most efficiently under a specific set of conditions and homeostasis maintains those conditions. For example, most enzymes in the human body work best at 37°C.

Mammals and birds can maintain a constant body temperature, but many other organisms cool down at night and, since their metabolic functions don't work as well at cooler temperatures, they are sluggish until they warm up.

Illness or extreme conditions can overwhelm the body's homeostasis.

11.1 Define this Key Term

homeostasis

11.1 Questions for Review

11.1.1 Explain what homeostasis means using two different examples.

11.1.2 Why is it important that your body temperature be maintained within a very narrow range?

11.2 How do Homeostatic Systems Work

11.2 Learning Objective:

Identify the three components of a homeostatic system, and explain the role of negative feedback.

11.2 Summary

Any homeostatic system consists of three parts:

1.   Sensor: This measures whatever property is being regulated, such as temperature or water in the blood.
2.   Effector: This has the ability to change the value of the system. A positive effector increases the value, while a negative effector decreases the value. An example of this in the body would be drinking *versus* urination.
3.   Controller: This registers information from the sensors and sends signals to the appropriate effectors to tell them what to do. This also establishes the set point, or the normal value for the system (for example 37°C body temperature in a human).

Homeostasis is maintained by negative feedback. If the system is above the set point, make it lower. If it's below, make it higher. Here is how it works:

1.   The controller reads the signal from the sensor.
2.   Based on the reading, the controller does one of three things:
     a.   If the signal is above the set point, it activates the negative effector.
     b.   If the signal is at the set point it does nothing.
     c.   If the signal is below the set point, it activates the positive effector.

11.2 Define these Key Terms

sensor

effector

controller

set point

11.2 Question For Review

11.2.1 Draw a flow chart illustrating how homeostasis is maintained by the three parts of the system.

11.3 How Does Your Body Sense Temperature?

11.3 Learning Objective:

Contrast the role of core and skin temperature sensors.

11.3 Summary

There are two ways the body determines its temperature.

1. Core Temperature Sensors

The hypothalamus is a region near the base of the brain, just above the brain stem. It connects with the pituitary gland that secretes many types of hormones, the chemicals that regulate the organs of the body. Therefore, the hypothalamus regulates homeostasis for many of the body's systems.

If the core body temperature increases or decreases, the blood temperature follows along, and inside the hypothalamus are nerve cells bathed in blood. Those nerve cells are the core temperature sensors. If the blood temperature decreases, the chemical reactions in the nerve cells slow down. If the blood temperature increases, those chemical reactions speed up.

2.   Skin Temperature Sensors

Since your skin is in contact with the external environment, nerve cells throughout the skin detect environmental conditions, including temperature. Different sets of nerves detect warmth and cold and send that information to your brain in sort of an advanced warning system. This allows your body to make adjustments to the external environment before changes occur in the internal environment, minimizing changes in body temperature.

11.3 Define these Key Terms

skin temperature sensors

core temperature sensors

hypothalamus

11.3 Questions for Review

11.3.1 What are the two ways the brain detects temperature?

11.3.2 Why are the skin temperature sensors especially important in maintaining body temperature?

11.4 How Does Your Body Adjust Temperature

11.4 Learning Objective:

Explain how the body uses effectors to change body temperature.

11.4 Summary

Your body is constantly at work, and that work generates heat like a furnace. It's pretty easy to increase the heat by increasing the activity, but it's much more difficult to get rid of excess heat. This is why even a small increase in temperature can be dangerous or fatal.

All of your ten trillion cells constantly conduct chemical reactions, which generate heat. Collectively, those chemical reactions are your metabolism. Metabolism is measured by how much heat in Calories you use.

Your basal metabolic rate is the number of Calories you need to maintain your body's processes at rest. This generates enough heat to maintain normal temperature. Whenever your body does work---digesting food, physical activity, thinking---your metabolic rate increases above the basal rate and that excess heat must be expelled.

Effectors that Cool the Body

Heat flows from a warmer object to a cooler one by convection. Because our bodies are usually warmer than the environment, we tend to lose heat through our external surfaces, mainly skin and lungs.

1. Blood circulation is one cooling effector. Blood warms up areas of high metabolic activity, like muscles or active organs, and transfers this heat to cooler regions. One of the body's cooling effectors directs warm blood to the skin where it comes into close contact with the cooler external environment and can lose heat to the cooler air. The bigger the difference between the object and the fluid, the more heat is transferred. Also more heat is transferred if the fluid is moving faster.

2. A second cooling effector is sweating which takes advantage of evaporation. Heat is required to evaporate the water in sweat, and this heat is taken from our body. An adult has 2-3 million sweat glands concentrated in areas of the skin that have good blood flow and tend to stay warm. So the blood effector and the sweat effector work together.

3. When you exhale, water evaporates from the moist surfaces of the lungs and heat is lost to the external environment.

On a hot day, when air temperature approaches body temperature, the rate of heat loss will be low. If it's also humid, sweating becomes less effective. You lose water but it doesn't evaporate and so you dehydrate. Heat stroke, when the body heats up faster than the effectors can cool it down, is the result.

Effectors that Warm the Body

1. Shivering warms the body by increasing the metabolic rate. Those muscle contractions increase the heat the muscles produce, and that heat is used to warm you.
2. Blood circulation changes can maintain heat. If the blood vessels near your skin and extremities are partially closed down, less blood flows to your skin and your skin temperature drops, closer to that of the external environment, reducing convective heat loss. The body's core stays warm.
3. Goose bumps are a warming effector that no longer work for humans. They are caused by the contraction of muscles near hair follicles, which makes hair stand up. This deeper hair or fur traps air pockets near the skin, and because the air is not moving, heat loss through convection is reduced.

Insulation is not considered a warming effector because it can't be turned on and off, but it helps prevent heat loss by radiation. Heat loss by radiation of infrared rays (what night vision goggles "see") goes on all the time, regardless of the external temperature. Fur and feathers provide insulation, and mammals also have a layer of fat just below the skin which acts as insulation. This is one of the key reasons mammals can maintain a constant body temperature. Snakes, lizards, frogs and fish all produce metabolic heat, but they have no way to maintain it. Fat can also be used to produce energy to raise the metabolic rate and warm the body.

11.4 Define these Key Terms

metabolism

basal metabolic rate

convection

evaporation

humidity

heat stroke

radiation

insulation

11.4 Questions for Review

11.4.1 What is metabolism?

11.4.2 Fill in the following table:

Cooling Effector	How it cools the body
Blood circulation	
Sweating	
Exhalation	
Warming Effector	**How it warms the body**
Blood circulation	
Shivering	
Goose bumps	

11.4.3 Explain why you feel hotter on a humid day.

11.4.4 What is heat stroke?

11.4.5 What is insulation and why is it important in maintaining temperature?

11.5 How Does Your Body Maintain a Constant Temperature?

11.5 Learning Objective:

Assess how sensors in the hypothalamus, fever, and behavior help your body maintain a constant temperature.

11.5 Summary

The temperature control region of the hypothalamus receives constant input from its own temperature sensors and temperature sensors in the skin, and it controls the set point (98.6°F or 37°C in humans). It compares the sensor input to the set point and directs the appropriate warming or cooling effectors. If the body is too warm, blood flow is diverted toward the skin and sweating begins. If it's too cool, blood flow to the skin is reduced and shivering occurs. Skin sensors allow it to anticipate change and react proactively, preventing a change in body temperature.

Fever

A fever is a temporary increase in body temperature due to infection. The higher temperature may be harmful to the parasites, but not to us, unless it's too high.

When your immune system detects and infection, certain blood cells produce a chemical signal that binds to cells in the hypothalamus, raising the set point temperature. The hypothalamus then triggers shivering, chills, and sweating, as it adjusts the body to the new set point.

Conscious Behavior

Physiological homeostatic processes work on their own, but you are aware of feeling hot or cold, so your behavior helps control your temperature. You can take advantage of the various mechanisms of heat transfer: convection, evaporation, and radiation.

You use convection when you move into the shade on a hot day, or inside on a cold one. If you increase the difference in temperature between your body and the external environment, you will lose heat more slowly. If you increase it, you will lose heat more quickly. You can control heat loss through radiation by putting on or taking off clothes. Clothes act as insulation.

Conduction is one final form of heat transfer. Heat flows from warmer to cooler in this case, but unlike with convection, in conduction, both objects are solids and neither moves in relation to the other. Ice packs and heating pads are good examples of conduction.

It's important to note that while birds and mammals can maintain constant body temperatures through physiology and behavior, other animals have lower metabolic rates and less insulation so they have a harder time staying warm. They do use behavior to keep their body temperatures within reasonable limits, however.

11.5 Define these Key Terms

fever

conduction

11.5 Questions for Review

11.5.1 You're playing tennis on a hot day. Describe how your body is helping you to maintain your temperature. (Don't forget that you behavior matters too.)

11.5.2 You're walking on a windy day and it's starting to snow. Describe how your body can help you maintain your temperature.

11.5.3 What is a fever? What triggers it? How is it helpful?

11.5.4 What is the difference between conduction and convection?

11.6 How Does Your Body Regulate its Fluids?

11.6 Learning Objective:

Describe how the body uses osmosis and diffusion to regulate its fluids.

11.6 Summary

Life first evolved in the sea and so dehydration wasn't much of a problem. On land, however, the cellular environment must stay moist despite evaporation. There are a few key words to know.

1. Diffusion: Our cellular environment consists mainly of molecules dissolved in water. Molecules are always moving, and they will spread from an area where there are lots of them to an area where there are fewer: down a concentration gradient. This movement from high concentration to low is called diffusion.

2. Permeability: Cells are surrounded by membranes that separate them from the fluid-filled spaces between cells. Some molecules can pass through those membranes and some can't. If a molecule can pass through, the membrane is permeable to that molecule. Most membranes are selectively permeable.

Permeability depends on the molecule's size and its electrical charge. Some dissolve in the membrane and cross easily. If cells need a particular molecule it may have proteins on the surface that form pores to allow the molecules to pass through, or pull the molecules in or out. This is active transport which requires energy.

If a cell is permeable to a molecule, it will passively diffuse down the gradient into or out of the cell until the concentration is equal on both sides of the membrane. If the cell needs to move a molecule against the gradient, it will use active transport.

3. Osmosis: This is the diffusion of water through a semi-permeable membrane from an area where dissolved molecules (solutes) are in low concentration to where they are in higher concentration. This means that if the fluid external to a cell is too highly concentrated, water will leave the cell and it will shrivel and die. If the contents of the cell are too concentrated, water will move into the cell and it will burst.

The organ that controls this important fluid balance is the kidney.

11.6 Define these Key Terms

diffusion

osmosis

permeability

active transport

osmosis

11.6 Questions for Review

11.6.1 Draw a diagram of a cell and illustrate the process of diffusion of a molecule to which the membrane is permeable.

11.6.2 Illustrate the process of active transport on your cell diagram. When would this be necessary?

11.6.2 For each of the following cases, consider which direction water would move relative to the cells involved:

a. A lettuce leaf is soaked in salad dressing.

b. You sprinkle salt on a slug (a shell-less terrestrial snail).

c. A piece of potato is soaked in salt water.

d. A piece of sea weedis soaked in salt water.

11.7 What Does the Kidney Do?

11.7 Learning Objective 5.7:

Outline the steps involved in making urine.

11.7 Summary

The main job of our two kidneys is to make urine, and in the process, to control water loss and maintain the proper concentration of many dissolved molecules that circulate in the blood.

Urine consists of water and waste material from the chemical reactions of metabolism. Nitrogen is the most common waste material, usually in the form of ammonia. Your liver converts ammonia to urea, and your kidneys remove that from the blood and put it in urine for elimination.

The Structure of the Kidney

The outer edge of the kidney is the cortex. The inner region is the medulla, and the tissue fluid there is very salty. In the middle is a hollow cavity called the pelvis where urine produced by the cortex and medulla collects before being sent to the urinary bladder.

The cortex is composed of about one million long thin tubules called nephrons. Each one begins with an open bulb called Bowman's capsule. In there is a network of small blood vessels called the glomerulus. The entire nephron is enmeshed in a capillary bed, a network of small blood vessels.

The Function of the Kidney

The long tubules of the nephron wind through the cortex and medulla, and the conditions outside the nephron change, allowing the nephron to do its job via diffusion, osmosis, and active transport. There are three parts of the process:

1. Filtration: In filtration, blood is pumped into the vessels of the glomerulus under somewhat high pressure which forces blood out of the pores of the vessels. The blood cells don't fit out, so the filtrate (the portion that does get through the pores) is about 20% of the fluid and includes salts, sugars, urea and amino acids.
2. Reabsorption: Some of those compounds that get filtered out are needed by the body so they get reabsorbed into the tissues of the kidney. Some of those then diffuse back into the blood vessels.
3. Secretion: This is the opposite of reabsorption. The body doesn't need some of the compounds that don't get filtered out, so these diffuse out of the blood vessels and are grabbed by the nephron in active transport and moved to the filtrate.

How the Kidney Makes Urine

The first tubule of the nephron reabsorbs most of the water, salt, potassium, amino acids, and other useful compounds, and returns them to the bloodstream. The second tubule removes even more water but retains urea, which results in a very concentrated urine. Salt is added to it by active transport, and more water removed by osmosis as it travels through the loop of Henle. From there it moves to the collecting tubules which merge into the collecting ducts which empty into the kidney's pelvis, and from there to the urinary bladder.

11.7 Define these Key Terms

urea

cortex

medulla

nephron

filtration

reabsorption

secretion

11.7 Question for Review

11.7.1 Using either the diagram or draw a new flow chart, describe the production of urine as it proceeds through the kidney. Indicate where filtration, reabsorption, and secretion occur, and where osmosis and active transport come into play.

11.8 How Does the Kidney Maintain Water Balance?

11.8 Learning Objective:

Predict the impact of drugs and the environment on water homeostasis.

11.8 Summary

The kidney is part of the homeostatic system that maintains both water and salt balance in the body.

How Water Enters and Leaves the Body

Humans get some water from food and some as a product of the chemical reactions of metabolism, but mostly we drink it.

We lose water when we exhale, and it evaporates throughout skin. We lose some through normal feces and a lot through diarrhea (this is a major cause of death in developing countries).

We can lose up to a liter of water per hour through sweat, although not normally. If you sweat too much, you dehydrate and can't sweat. Heat stroke is likely to result and this is a significant cause of kidney failure. In addition, the dehydration causes blood pressure to drop so the heart has to work harder, creating even more heat.

 Finally we also lose water through urine, about 1.5 liters per day. The kidney can adjust this as needed.

Osmotic Sensors

Blood is the indicator of water in the body. The sensors are specialized nerve cells in the hypothalamus. If blood is too concentrated, water flows out of the nerve cells by osmosis and they shrivel and fire, indicating that water needs to be retained.

The Water Balance Effector

The water balance effector is the antidiuretic hormone (ADH). It reduces the volume of the urine produced, reducing water loss. ADH is produced by the pituitary gland in the brain. It is released into the blood stream and binds to cells in the collecting tubules of the kidney. There special pores open to allow water to move across the wall of the collecting tubule and back into the blood, decreasing the volume of urine.

The Water Balance Controller

The hypothalamus controls homeostasis for water balance. After ADH is released and water is retained, the blood becomes less concentrated. The osmotic sensor nerves swell and stop firing and the hypothalamus stops signaling the pituitary gland to release ADH. Water balance is achieved.

Disrupting Water Balance

Drugs can disrupt water balance. Alcohol, for example, is a diuretic, meaning it causes you to lose water (the opposite of the anti-diuretic hormone). It interferes with ADH binding. Dehydration is one of the contributors to the "hangover." There are other drugs that can do this. Too much salt in your diet can also cause you to lose too much water.

11.8 Define these Key Terms

antidiuretic hormone

diuretic

pituitary gland

11.8 Questions for Review

11.8.1 How does the hypothalamus detect dehydration?

11.8.2 Describe how ADH aids in preventing dehydration.

11.8.3 How is a drug like alcohol a diuretic?

11. Biology in Perspective

The body's internal balance is finely tuned to maximize the efficiency of chemical processes, and is maintained by a process called homeostasis. The hypothalamus in the brain regulates this. Under extreme conditions though, such as extreme temperature conditions or dehydration, homeostatic control can fail and this has severe consequences.

Question to consider: Why is it important that the body be kept in homeostatic balance?

11. Scientist Spotlight: Homer William Smith

How did Homer William Smith's focus on the scientific method influence the area of kidney disease research?

For 30 years, Dr. Homer William Smithworked on kidneys. He developed noninvasive ways to study the kidney in living organisms, and he figured out the details of how the kidney filters, reabsorbs, and secretes compounds.Smith worked to bridge the gap between laboratory research and medical research. This brought both scientific and medical research to the problem of kidney disease and led to improvements in treatments and understanding of kidney health,

Question to consider:

What did Homer William Smith do in the field of physiology that led to great advancements in the understanding of kidney function and disease?

11. Technology Connection: Dialysis

How is it possible for someone to live without functional kidneys?

Smoking, drinking, disease and chemotherapy drugs can all damage a kidney. You have two, but you need at least one to work. If your kidneys fail, you need dialysis. In dialysis, blood is circulated outside the body through an artificial kidney or dialysis machine. Chemicals are added so the blood doesn't clot, and the machine has filters to eliminate air bubbles that could cause a stroke. The machine maintains the right amount of water in the blood, removes metabolic wastes, and maintains the right concentrations of various dissolved chemicals like salt. The blood passes in thin tubes through a fluid specifically blended for the patient. The whole process then works by diffusion and osmosis.

Question to consider: Generally, what does a dialysis machine do and how does it do it?

11. Life Application: How to avoid dehydration

What role did the University of Florida play in the creation of Gatorade?

Exercising in hot humid conditions can leads to loss of water and also a host of other dissolved chemicals: salt, potassium, calcium, magnesium, and some trace metals. This upsets the osmotic balance of your bodily fluids, which stresses your cells and leads to a feeling of being run down and exhausted. Competing athletes can't afford these symptoms, so in 1965, coaches for the University of Florida football team turned to scientists who quickly realized that water alone wasn't enough in these extreme conditions. They developed a liquid that was essentially artificial sweat, designed to replace not onlythe waterbut also salts and carbohydrates the athletes were losing. It tasted really bad, so they added lemon juice to make it a little bit tastier.

The team went 7-4 in 1965 and 9-2 in 1666 and won the Orange Bowl. The team is the Florida Gators, so their drink was named Gatorade.

Question to consider:

How is drinking Gatorade a bit more helpful in maintaining fluid balance than just drinking water? Is it surprising this was developed in Florida? Why or why not?

5. How Do We Know?:Hypothalamic Control of Temperature

How did scientists use dogs to show that the hypothalamus can sense changes in body temperature?

In the 1960s, Dr. H. T. Hammel and his lab groups performed experiments to see how dogs controlled their body temperature. They inserted thin tubes into the dog's brains. The tubes ended at various points around the hypothalamus. The tube was closed at the end in the brain and open outside the skull so the scientists could add things. The dogs recovered from this well, and the experiments commenced.

The dogs were placed in a "neutral" thermal environment: neither hot nor cold (a dog's body temperature is a little warmer than ours—101°F or 38.3°C. Hammel introduced fluid into the tubes. If the fluid was cool, the dogs began to shiver and the skin blood vessels constricted. If the fluid was warm, the dogs panted and their blood vessels dilated. In both cases these effectors changed the body temperature of the dogs, but the hypothalamus continued to trigger them as long as the temperature around it (in the tubes) stayed warm or cool.

This showed that the hypothalamus could sense changes in temperature and it responded to warm or cool the body as needed. Because the room temperature stayed neutral, it was clear that there were no temperature signals coming from the skin.

Question to consider:

Outline the experiment Dr. Hammel did. How did it demonstrate that the hypothalamus senses and responds to body temperature changes?

Student Study Guide

Chapter 12 Circulation and Respiration

What If Your Body Doesn't Get the Oxygen It Needs?

Most people can't hold their breath for more than a couple minutes. We can't store much oxygen and our cells need it continuously to make energy. Every chemical reaction in your body ultimately depends on oxygen. Metabolically active cells shut down without oxygen, and if your heart or brain loses oxygen for even a few minutes, the consequences can be fatal.

Your circulatory system carries blood through the body, delivering oxygen you obtain through your respiratory system. These systems are controlled by your brain, which monitors oxygen levels with sensors throughout the body. This system works well, but it does have its limits.

12. Case Study: Blood Doping at the Tour de France

Why would elite athletes risk their careers and their health to engage in blood doping?

12. Case Study Summary

Blood doping is a prohibited and risky process that enables your blood to carry more oxygen, giving athletes more endurance. First the athlete "donates" a pint or two of his or her blood. This is placed in a tube in a centrifuge, which spins around so fast that the red blood cells---the ones that carry the oxygen---get forced to the bottom of the tube. The rest is injected back into the athlete and red blood cells are stored.

A day or so before a competition, the stored red blood cells are injected back into the athlete. These extra cells allow the blood to carry more oxygen. Although it's prohibited, bike racer Lance Armstrong said in a 2013 interview that he could not have won the *Tour de France* 7 times if he had not done this.

Question for Review

How does blood doping work to increase a body's oxygen capacity?

12.1 How much Oxygen do you Need?

12.1 Learning Objective:

Explain how oxygen needs vary depending on circumstances.

12.1   Summary

At rest an adult needs between 0.2 and 0.3 liters of oxygen per minute. During active exercise you might need 10 times that, or 2 to 3 liters per minute. World-class athletes like those in the Tour de France may increase their oxygen demand 30 times the resting rate to between 6 and 9 liters of oxygen per minute.

How much oxygen can you take in?

How fast you can get oxygen depends on how much oxygen is in the air and how quickly your blood circulates. The air we breathe is 21% oxygen. Each breath takes in 0.5 liters of air or 0.1 liters of oxygen. At rest, an adult typically breathes about 12 times per minute, so about 1.2 liters of oxygen enters your lungs each minute at rest. In strenuous exercise, an adult might breathe 30 times per minute and take in about 2 liters of air per breath. This provides about 12 liters of oxygen per minute. A world-class athlete might breathe 60 times per minute, taking in 24 liters of oxygen per minute.

How fast can your blood circulate?

Your heart at rest typically beats about 60 times per minute and pumps about 5 liters of blood in that time, which is about all the blood you have. When you are more active, your heart rate speeds up and your circulation is faster. This would provide plenty of oxygen. Unfortunately though, not all the oxygen you breathe in makes it from the lungs to the bloodstream. Oxygen doesn't dissolve well in water, or blood. A liter of blood can only carry about 5 milliliters of oxygen, which isn't enough for a body even at rest. To get around this problem, the red blood cells contain the protein hemoglobin, which holds onto oxygen and transports it around the body. With this, normally a liter of blood can hold 200 milliliters of oxygen, which means you have plenty of oxygen available.

What's the limit?

The more active you are, the more oxygen gets to your muscles via your breathing and circulation, but at some point you start running out of breath. You don't have enough red blood cells to carry the extra oxygen you need. One thing training does is increase your number of red blood cells to increase your oxygen capacity. Blood doping does this artificially.

12.1 Define this Key Term

hemoglobin

12.1 Questions for Review

12.1.1 Fill in the following table comparing oxygen intake in theory.

	Breaths /min	O2 per breath	Liters O2 needed/min	Liters O2 available per minute
Adult at rest				
Active adult				
Active athlete				

12.1.2 According to the table above, even an athlete should have plenty of oxygen. Why don't they? What's limiting it?

12.1.3 How does the body ensure it typically has enough oxygen even though oxygen doesn't readily dissolve in blood?

12.2 How is the Circulatory System Structured?

12.2 Learning Objective:

Describe the role of the different components of the circulatory system.

12.2 Summary

The circulatory system is a series of vessels. Larger vessels branch into ever smaller and more numerous ones such that they can reach every cell to provide nutrients and oxygen and remove wastes. This type of branching is common in biological structures and increases surface area. In the circulatory system, this allows efficient transport and distribution of oxygen and other chemicals. Materials travel from the heart to arteries, capillaries (the smallest vessels), back to veins, and then the heart.

The Heart

Your heart beats around 60 times per minute, more when you were younger or are more active, for your entire life. It is composed of two biological pumps side by side. Each atrium is a hollow chamber with weakly muscular walls. The atrium receives the circulating fluid and acts as a reservoir that fills the ventricle. The ventricle has strong muscular walls that force fluid along when they contract. The right-hand ventricle receives "used" blood from the body and sends it via the pulmonary circuit to the lungs to reoxygenate. The left-hand ventricle gets blood from the lungs and sends it via the systemic circuit to the rest of the body.

A series of valves in the heart keeps blood flowing through the four chambers in the correct direction. The atrioventricular valve lies between the atrium and its corresponding ventricle. When the atrium contracts the valve opens into the ventricle, but when the ventricle contracts, the valve closes so blood doesn't go back into the atrium. A semilunar valve lies between each ventricle and the blood vessel it uses to send blood from the heart.

When blood doping is done, the extra red blood cells thicken the blood, forcing the heart to work harder to pump the thicker fluid. This increases the risk of a heart attack, and may also reduce the circulation rate.

Arteries

Arteries are the blood vessels that transport blood away from the heart. They have thicker walls to withstand the pressure of the contraction of the ventricles. The inner layer of the arterial wall is smooth to minimize friction. The middle layer contains a band of muscle cells and elastic fibers. The muscles help to control blood pressure, but they are involuntary muscles meaning they are controlled by the blood flow itself, and hormones. The outer layer is mostly a thick, fibrous coating that protects against piercing and damage.

The arteries start with the biggest artery leaving the heart, the aorta, and they branch throughout the body. The smallest are called arterioles. These then branch into the even smaller and more numerous capillaries.

Capillary Beds

Arteries transport blood and capillary beds distribute the nutrients and chemical signals in the blood to the cells. They also remove carbon dioxide and wastes. They are found everywhere in the body since all the cells need this service.

Where arterioles branch into capillaries, there are small bands of involuntary muscles called sphincters. These can be opened or closed to direct circulation toward or away from a particular capillary bed as needed. At any given time, most of your capillary beds are closed, and this adjusts as your activities change over the day. This opening and closing of beds is microcirculation. If all the beds opened at once, there would not be enough blood for the heart and circulation would stop. This life-threatening condition can be caused by an injury and is called hypotensive shock.

Veins

From the capillary beds, blood flows into small vessels called venules, which merge together into veins. These transport blood back to the heart. By the time blood gets to the veins, it has very little pressure, so veins lack the elastic fibers in arteries, but they have muscle structure because their contractions, along with the contractions of your skeletal muscles, help move the blood. Veins have a series of one-way valves to keep blood from flowing backwards.

Blood

Blood is made of liquid plasma and a variety of types of cells. Plasma is 90% water and makes up about 55% of blood. The remaining 10% is nutrients like proteins, amino acids, carbohydrates, and fats, hormones, metabolic wastes, chemicals from the immune system, dissolved gases like oxygen, carbon dioxide, and nitrogen, and some pollutants.

Blood cells make up 45% of you blood and 99% of these are the red blood cells that transport oxygen. The rest are platelets, which are fragments of blood cells responsible for blood clotting, and white blood cells that are part of your immune system.

Blood doping reverses the percentages such that the blood is 55% red blood cells by volume.

12.2 Define these Key Terms

atrium

ventricle

artery

capillary

vein

pulmonary circuit

systemic circuit

atrioventricular valve

semilunar valve

microcirculation

plasma

12.2 Questions for Review

12.2.1 How are arteries and veins structurally different and why is this since they both move blood.

12.2.2 What is included in blood and what function does each part serve?

12.3 How is the Respiratory System Structured?

12.3 Learning Objective:

Describe the role of the different components of the respiratory system.

12.3 Summary

The respiratory system also depends on a system of branching to increase surface area for gas exchange, but it is different from the circulatory system in that the branches lead to dead ends. The same branches that bring fresh air in, also send stale air out.

Lungs

Your lungs are two large sacs. The left one is divided into two compartments or lobes, and the right into three. Within the lobes is a branched network of airways starting with the trachea, or windpipe, leading deep into the lungs. The first branches from this are the bronchi. These branch into bronchioles that end in tiny sacs called alveoli, clustered like grapes and surrounded by capillary beds. This is where diffusion of oxygen and carbon dioxide occurs. All this branching increases surface area such that you have 1500 miles of airways and the surface area of the alveoli is about 70 square meters, or half a volleyball court. There is plenty of space for gas exchange, so it's not here where oxygen intake is limited. It's the ability of the blood to carry the oxygen and this is why extra red blood cells, preferably gained through training, are helpful for athletes.

The Respiratory Pump

The expansion and contraction of your lungs for inhalation and exhalation is accomplished by your ribs and the muscles attached to them, along with the diaphragm. This circular sheet of muscle separates your chest from your lower abdomen. The pump works by changing the volume of your chest cavity. When the cavity increases, your lungs expand to fill it and the expansion draws air in. When the chest compresses, the lungs compress as well, pushing air out.

When the diaphragm contracts it flattens out. This, along with muscles rotating the ribs upward and outward, expands the chest cavity for an inhalation. Other muscles in the ribs can also contract to allow more deep breathing more quickly.

Nose and Mouth

Air can get into and out of the trachea via the nose or the mouth. Your nose warms, moistens and filters the air you inhale. It also detects odors. You can also breathe through your mouth and because your nose and mouth are mostly separate, you can breathe and eat simultaneously.

Your nasal and oral passages come together at the back of your mouth, which is helpful when your nose is too stuffy from a cold to take in air. It also allows the possibility of choking, though, when your trachea gets blocked. The epiglottis is a flap of skin at the top of the trachea that covers it when you swallow, helping to prevent choking

12.3 Define these Key Terms

trachea

bronchi

alveoli

diaphragm

epiglottis

12.3 Questions for Review

12.3.1 Describe how the structure of the respiratory system allow for efficient gas exchange?

12.3.2. How does inhalation and exhalation happen?

12.3.3 What does a nose do as part of the respiratory system?

12.4 How Do You Breathe?

12.4 Learning Objective:

Explain how and where oxygen and carbon dioxide diffuse into and out of blood and the role of hemoglobin to facilitate oxygen transport.

12.4 Summary

Respiration involves taking air into the lungs, transporting the oxygen to every cell, and removing carbon dioxide. This is gas exchange.

Getting Oxygen

When you inhale, air rich in oxygen (21%) and low in carbon dioxide (0.04%) travels to your alveoli. These sit by capillary beds. The blood in these comes from the "used" side of the heart and so is very low in oxygen and high in carbon dioxide picked up from metabolizing cells. The gases can, therefore, simply diffuse down the gradient of concentration. Oxygen leaves the alveoli to enter the blood and carbon dioxide enters the alveoli from the blood. The oxygenated blood moves back to the heart and on to the body. The carbon dioxide-laden air moves into the lungs to be exhaled.

Carbon dioxide dissolves easily in water, or blood, so it diffuses easily. Oxygen does not, so it needs to be "grabbed" by the hemoglobin in the red blood cells, and taken out of the liquid portion of the blood so that more will diffuse in. You can see how having extra red blood cells would enhance this process, providing extra oxygen.

Transporting and Delivering Oxygen

Blood transports oxygen to any open capillary bed. This is where oxygen delivery occurs. The metabolizing cells use up their oxygen and produce carbon dioxide so again there is a concentration gradient between the blood in the capillary bed and the adjacent cells. The gases can diffuse down the gradient. The cells gain oxygen and the blood gains carbon dioxide. The capillary beds also deliver nutrients, hormones, and other chemical signals, and pick up other waste products.

The "used" blood is now depleted of oxygen, returns via the veins to the heart, and gets pumped back into the lungs to pick up more oxygen, and the process continues.

The Chemistry of Gas Exchange

Hemoglobin in red blood cells is composed of four protein molecules, each containing one atom of iron. Oxygen combines readily with iron to form iron oxide or rust. The red blood cells can then carry this oxygen around the body. Hemoglobin binds tightly when there is a lot of oxygen present, and loosely when there is less oxygen. In the lungs, where oxygen concentration is high, hemoglobin binds quickly and tightly to every molecule. When this hemoglobin arrives at the capillary beds, oxygen content is low, changing the chemical properties of the hemoglobin such that it releases its oxygen allowing it to diffuse to the cells that need it. This means the more oxygen-depleted a cell is, the faster it will be able to pick up oxygen. So the active muscle cells of an athlete could pick up even more oxygen if the blood had the extra red blood cells carrying it.

When carbon dioxide dissolves in the blood, it produces carbonic acid. Therefore, in active areas of the body which are using up lots of oxygen and producing lots of carbon dioxide, the blood will get increasingly acidic. The pH of the blood is a measure of this, and a low pH indicates that you need more oxygen. Sensors in the body detect this and cause the heart and lungs to pump and breathe faster.

12.4 Define this Key Term

gas exchange

12.4 Questions to Review

12.4.1 Explain how diffusion moves oxygen and carbon dioxide into and out of the blood stream in the lungs.

12.4.2 Now explain how the process you describe above reverses in the muscles.

12.4.3 How do the chemical properties of hemoglobin aid in getting sufficient oxygen to muscles? Why is it necessary?

12.4.4 How does the body "know" when it needs to increase its heart or breathing rate?

12.5 How Does Blood Pressure Vary Throughout Your Body?

12.5 Learning Objective:

Describe the differences in blood pressure in different positions in the circulatory system as well as the mechanisms that assure blood flow.

12.5 Summary

Blood leaving the heart to enter the systemic circuit (the body) is under considerable force known as blood pressure. This causes the arteries to stretch and forces blood towards the capillary beds. How strong the pressure is depends on what you are doing and where in the body you measure it. Blood pressure is lowest when you are at rest. It is highest in the left side of the heart when you are active, because this pressure must get the blood throughout the body. The right side of the heart provides blood to the lungs and so there the circuit is much shorter, and the lungs are too fragile for very high pressure. This is why it's helpful that your heart consists of two pumping systems.

Blood pressure drops as blood moves through the systemic circuit into the capillary beds. After slowly moving through the beds, the pressure in the venules is essentially zero; this is why the veins and skeletal muscles need to move blood in the veins by contracting. Friction causes some of this decline in pressure, but it mostly has to do with the fact that the capillary beds are so numerous that they hold a bigger volume of blood than the arteries. The blood, essentially, can spread out, slow down, and allow more time for diffusion to occur.

Blood pressure that is too high is dangerous because it can damage arterial walls, cause circulatory problems like atherosclerosis, or cause strokes.

12.5 Questions for Review

12.5.1 Why are the arteries built to withstand higher blood pressures than the veins?

12.5.2 Why is the blood pressure in the right side of the heart lower than in the left?

12.5.3 Why does blood pressure drop between the arteries and the veins?

12.6 How are the Respiratory and Circulatory Systems Controlled?

12.6 Learning Objective:

Assess the effect that changes in the blood and nervous system have on respiration, heart rate, and blood pressure.

12.6 Summary

How does your body "know" when it needs to speed up or slow down circulation and respiration, and how does it do it?

Control of Respiration

Respiration is part of homeostasis. The breathing centers in your brain are at the top of your spinal cord: the medulla oblongata and the pons. Nerves run from these to the muscles in your chest, ribs, and the diaphragm, triggering contraction for each inhalation, and relaxing in between for exhalation.

When your body needs oxygen, oxygen levels go down and carbon dioxide levels, or carbonic acid, goes up. If the medulla oblongata and pons detect lowering pH in the blood (increasing acidity), they will increase their nerve firing rate, making you breathe rapidly and deeply. If pH rises, firing is reduced and breathing slows and becomes shallow. There are also blood pH sensors in the aorta and carotid arteries, the two main arteries supplying blood to the head.

Control of Blood Pressure

Blood pressure increases if the heart beats faster or more strongly and decreases if the contraction rate or strength decreases, but a lot of other factors go into this.

Pressure sensors are found in the aorta and carotid arteries. These specialized nerve cells respond to the stretch in the artery as the heart beats. If the heart beats more quickly, the arteries stretch more, and blood pressure increases. This information is relayed to the medulla oblongata and pons that control the heart rate, and therefore the blood pressure, by negative feedback.

Control of the Heart Rate

Heart cells beat all by themselves, but to beat in a coordinated fashion, they need a controller called a pacemaker. This is located in the right atrium. When its cells contract, the electrical signal to contract goes to the atria, then on to the ventricles, then the cells relax before the pacemaker sends out the next signal. This process keeps the heart cells beating in a coordinated way, but many things control the heart rate.

The vagus nerve, from the medulla, connects to the pacemaker to slow it down: an inhibitor. A second group of nerves from your spinal cord speed up the pacemaker: an exciter. Your heart rate depends on the balance between these. It is also controlled by hormones. If you are nervous or excited, adrenaline released into your blood will increase your heart rate and blood pressure. Another chemical signal, acetylcholine, slows down the heart rate. The vagus nerve can trigger this. A lower than normal level of the hormones that speed up the heart would slow it down. This might happen if your thyroid gland is dysfunctional, which would then cause you to feel fatigued.

12.6 Define these Key Terms

medulla oblongata

pons

pacemaker

vagus nerve

adrenaline

acetylcholine

12.6 Questions for Review

12.6.1 Describe what happens to lead to an increase in your breathing rate when you are exercising.

12.6.2 Explain how control of the heart rate is an example of homeostasis.

12.6.3 What do adrenaline and acetylcholine do?

12.7 What Happens when the Circulatory and Respiratory Systems Malfunction?

12.7 Learning Objective:

Evaluate medical advice concerning the treatment and prevention of atherosclerosis, heart attacks, and asthma.

12.7 Summary

Atherosclerosis

Atherosclerosis begins with inflammation of the arterial wall. The immune system responds by sending white blood cells to the site, and platelets collect, forming fibrous blood clots. Fats like cholesterol may accumulate and calcium deposits may form. The resulting structure is a plaque, which grows and narrows that part of the artery. This increases blood pressure upstream from the plaque, and decreases oxygen delivery and blood flow downstream. The artery may become blocked.

High blood pressure is the most important indicator that you have or may develop artherosclerosis. High cholesterol contributes to plaque formation. Your genes contribute to this, but life choices are important as well.

Heart Attack

The heart receives blood from the aorta through the coronary arteries. These can become clogged with plaque, just like other arteries, reducing blood flow and oxygen. If heart muscles use up all their oxygen the cells will begin to die. This is a heart attack. Atherosclerosis in other arteries can also cause a heart attack because plaques can break off and a clot can lodge in a coronary artery.

Blood doping aggravates the risk of heart attack because the thicker blood causes more labored beating of the heart that can damage the arterial lining. Reintroduction of the stored blood can also introduce tiny clots which increase the risk of arterial blockage, heart attack and stroke. Any blood transfusion carries these risks, but if it's for a medical necessity, it is worth the risk.

Stroke

Your brain occupies just 2% of your body but requires 20% of your oxygen. A blocked artery to the brain starves brain cells of oxygen and they begin to die. This is a stroke. The consequences are variable, depending which part of the brain is affected. Mini-strokes may be undetectable. Speech, vision, or use of limbs may be affected, or the stroke can be fatal.

Asthma

Asthma is actually an inappropriate allergic response that constricts the airways, making breathing difficult. The irritant causes the smooth muscles of the bronchioles in the lungs to contract or spasm, reducing or closing the airway. This is aggravated by inflammation and mucus in the lining of the respiratory tract.

Asthma is incurable but can be treated by avoiding triggers and use of drugs to open constricted airways. Its incidence has skyrocketed in industrialized nations lately. The reasons are not entirely understood but it may be that reduced exposure to infection actually increases the chance of developing asthma, possibly due to an "untrained" immune system that consequently overreacts to minor irritants.

12.7 Define these Key Terms

atherosclerosis

plaque

heart attack

stroke

asthma

12.7 Questions for Review

12.7.1 Describe what happens in atherosclerosis.

12.7.2 How does atherosclerosis potentially lead to both heart attacks and strokes?

12.7.3 What is asthma?

12. Biology in Perspective

Under most conditions, your body gets the oxygen it needs through the respiratory and circulatory systems. Both systems operate with pumps and numerous branches that increase surface area to allow efficient diffusion of gases. Red blood cells packed with hemoglobin allow for increased transport of oxygen. Sensors in your brain continuously monitor heart rate, respiration rate, blood pressure, and oxygen levels (through blood pH), and the brain uses negative feedback to adjust respiration and circulation accordingly.

If certain parts of the body are deprived of oxygen, atherosclerosis, blood clots, a heart attack, or stroke may occur. To some extent these risks are genetic, but lifestyle choices matter.

Question to consider:

Describe how both the respiration system and circulatory system use pumps and surface area to accomplish their functions.

12. Scientist Spotlight: Michael Ellis DeBakey

How did Dr. Michael DeBakey's invention save his own life?

While Michael Ellis DeBakey was still in medical school, he invented a roller pump that gently pushed fluid through a tube compressed between two rolling pins. His pump was incorporated into cardiopulmonary bypass machines because it circulated blood more gently than any other pumps at the time. DeBakey also developed several types of artificial hearts, stretchy Dacron tubes used to replace damaged arteries, and became one of the first surgeons to successfully perform cardiac bypass, artery replacement, and heart transplants.

In 1932 he volunteered for service in World War II. It was his idea to put Mobile Army Surgical Hospitals (M*A*S*H) right on the front, rather than having to move wounded soldiers. He also set up what would later become the Veteran's Administration.

Question to consider:

Describe a few of the contributions of Dr. DeBakey to medical science.

12. Technology Connection: Cardiopulmonary Bypass Machine

How is it possible to operate on the heart muscle when it continuously contracts?

It's nearly impossible to perform surgery on a beating heart, and the heart can't be stopped for surgery because cells in the heart and brain begin to die. The cardiopulmonary bypass machine, or heart-lung machine, has made these types of operations possible, safer, and much more common. The machine is connected to the patient's circulatory system and takes over the function of both the heart and lungs. The heart can then be stopped for surgery.

Question to consider:

Why is a cardiopulmonary bypass machine necessary for heart surgery, and what does it do?

12. Life Application: How Erections are Produced

Why did a doctor drop his pants and reveal an erection to a room full of distinguished scientists and doctors at a national medical conference?

When blood enters the penis through two major arteries, it flows into three sponge-like chambers. The muscles in these arteries are normally partially contracted so blood enters at a moderate rate. If these muscles relax, the artery dilates and blood enters at a very high rate, causing the chambers to expand, producing an erection. Normally this job is handled by the brain but an artificially introduced muscle relaxant works too. This means that erectile dysfunction (ED) can be treated with drugs.

One of the main causes of ED in older men is that their bodies don't produce enough muscle relaxant. In addition there is an enxyme, phsphodiesterase (PDE) that breaks down the relaxant to bring the erection to a natural end. Modern ED drugs prevent the PDE from working.

Question to consider:

Explain the physiology behind getting and maintaining an erection.

5. How do we know: Measuring Blood Pressure

What do the two numbers in a blood pressure reading mean?

Blood pressure is one of the four "vital signs" (along with pulse, respiration rate, and body temperature) that health professionals take to monitor your body's physiological state. Both the heart and the major arteries contribute to blood pressure which is why you have two numbers like 120 over 80. The first number is systolic pressure, when the heart contracts. The heart then relaxes and the arteries which were stretched by that pulse of blood, contract back to their normal size. This recoil is diastolic pressure, the lower number. Your blood pressure alternates smoothly between these two with every heartbeat.

To measure blood pressure, a cuff is inflated over your arm above the elbow and a medical professional listens with a stethoscope for the starting (systolic pressure) and stopping (diastolic pressure) of blood flow as the cuff is deflates. Pressure is measured in millimeters of mercury. A pressure of 120 represents enough pressure to hold up a column of mercury 120 mm high.

Question to consider:

If the doctor tells you your blood pressure is 120 over 80, what does that describe?

Student Study Guide

Chapter 13: The Nervous System

Does Your Brain Determine Who You Are?

What does it mean to be human? Is it all in the brain? Understanding the biological basis of human thought, behavior, and emotion is one of the biggest challenges facing biologists. Alzheimer's disease and other memory disorders, strokes, and injuries can cause irreparable damage cognitive and motor systems. Better understanding of how we think and move would likely lead to better methods of repairing these systems, as well as a better understanding of how our biology makes us who we are.

13. Case Study: The Only Living Man With a Hole in His Head

> *How could a man survive with a hole in his brain?*

13. Case Study Summary

In 1848, Phineas Gage, a railroad worker, was blasting through boulders for rails in Vermont. His job was to tamp down sand on top of gunpowder in rods that then went into holes drilled in the rock. Then the fuse was lit and the explosion shattered the rock. One day the sand didn't get poured, and when Gage tamped down with a metal tamping iron, there was a spark. The explosion sent the 3'7" rod right through Gage's head. Oddly, he sat up and started talking, despite a 3.5" wide hole in his head. It had entered just under his left cheekbone and passed behind his left eye, exiting just above the hairline toward the middle of the forehead.

Dr. John Harlow treated Gage. He successfully patched up the wound and eventually declared it "healed," and Gage went back to work. But Gage's personality had entirely changed. He had been responsible, hardworking, and dependable. He became unreliable and nasty, and he couldn't make decisions. Eventually, he was fired. He was healthy, but he was different. What happened?

Question for Review:

In your own words, explain what happened to Phineas Gage. What was the result?

13.1 What are the Parts of the Human Brain, and What Do They Do?

Learning Objective 13.1

> *Identify the parts of the brain and predict the effect of damage to those regions.*

13.1   Summary

At the time of the injury to Phineas Gage, physicians thought that either the brain was one functional unit or that it was composed of several independent activity centers. The fact that Gage remained fully functional after such a dramatic brain injury suggested that the brain was made of several separate activity centers. In fact, both views are accurate. So what did happen to Gage?

A brain weighs about 3 pounds. It's protected by the skull bones and the meninges, which is a tough connective tissue covering the brain. The brain consists of several subsections which usually work together:

1. Hindbrain

This is located at the base of the skull, right above the spinal cord. It controls many basic body functions and has three subunits:

- o Medulla oblongata: Influences sleep and waking, respiration and circulation, and coordinates the motor responses of reflexes like coughing and sneezing.
- o Cerebellum: Integrates what we see, hear, and feel; coordinates movement and balance; and has important role in manual dexterity and language.
- o Pons: Directs communication between the cerebellum and forebrain and helps control breathing and circulation.

2. Midbrain

The medulla oblongata, the pons, and the midbrain form the brainstem. All information moving to and from other brain regions passes through the brainstem. The midbrain has a key role in coordinating responses to light and sound.

3. Forebrain

This is the most highly developed and largest region of the brain and is composed of four parts:

- o Olfactory bulbs: Interpret sensory information about smell.
- o Thalamus: Relay switch to sort incoming and outgoing data.
- o Hypothalamus: Master control center regulates body temperature, blood pressure, hunger, sex drive, emotions, thirst, and physiological reactions to stress.
- o Cerebrum: Largest and most sophisticated part; 80–85% of brain mass in humans and is responsible for much of what makes us human: reasoning, mathematical and artistic ability, imagination, language, personality, and perception.

The Cerebrum

The wrinkled surface of the cerebrum is the cerebral cortex. It consists of a left hemisphere specializing in language, logic, and math, and a right hemisphere specializing in emotions, intuition, and artistic impression. The thick band of nerve cells called the corpus callosum connects the two hemispheres so they're in constant communication. Each hemisphere consists of four lobes:

- o Frontal: Speaking, problem solving, personality, motor control.
- o Temporal: Perception and processing information from eyes and ears, language comprehension, and pattern recognition, including faces.
- o Parietal: Perception and processing of physical touch, and coordinating perception from other senses.
- o Occipital: Perception and processing of information from the eyes.

Limbic System

The limbic system is part of the forebrain and includes parts of the cerebrum, the hippocampus, and the amygdala. It controls physical drive and instincts and is partly responsible for emotions, learning, and memory.

13.1 Define these Key Terms

hindbrain

midbrain

forebrain

medulla oblongata

pons

cerebellum

olfactory bulbs

thalamus

hypothalamus

cerebrum

cerebral cortex

corpus callosum

limbic system

13.1 Question for Review

13.1.1 Summarize the function of each of the following brain parts.

Brain Part	Function
Hindbrain	
Midbrain	
forebrain	
Medulla oblongata	
pons	
Cerebellum	
Olfactory bulbs	
Thalamus	
hypothalamus	
cerebrum	
Cerebral cortex	
Corpus callosum	
Limbic system	

13.2 What Was Wrong with Phineas Gage's Brain?

Learning Objective 13.2:

Assess how brain injury can alter personality using Phineas Gage as an example.

13.2 Summary

Following his injury, Phineas Gage was fully functional and went back to work, retaining his physical abilities and his dexterity, but his personality had changed. So what happened in his brain?

In 1994, scientists used computer imaging techniques to precisely map the trajectory of the rod through Gage's brain. It missed the motor and sensory strips on top of the cerebrum, so he was able to keep his balance, remember, and pay attention. It missed the Broca's area in the left temporal lobe, which is where language is produced. The main damage was in the frontal lobe in the left hemisphere. Patients with damage here do well in logic and math but can't make decisions about personal or social things. They are unable to "read" people, exhibit little empathy, and don't know how to be social.

13.2 Questions for Review

13.2.1 On the brain diagram, indicate the path of the rod in Gage's head.

13.2.2 Describe what the rod injured and what the consequences of that type of injury are for a person.

13.3 How Is the Nervous System Organized and What Does It Do?

Learning Objective 13.3:

Differentiate the roles of the central nervous system and the peripheral nervous system in controlling our actions, behaviors, and body functions.

13.3 Summary

About 10 years after the injury, Phineas Gage began experiencing epileptic seizures, sort of electrical storms in the brain cells which lead to convulsions. Gradually, these became almost continuous, and in 1860, at 38 years of age, Gage died. What happened?

Organization of the Nervous System

The nervous system is composed of signaling cells called neurons. The brain has more than 10 billion neurons, and the rest of the nervous system has more. Each can communicate with at least 1000 other neurons. This networking allows nearly unlimited combinations of signals that allow us to think, walk, speak, and feel. How is all this communication potential directed so it can be useful?

The Central Nervous System

Nerves are bundles of neurons and the spinal cord is a bundle of nerve fibers in our backbone that connects the brain to the rest of the body. The brain and the spinal cord comprise the Central Nervous System (CNS) that controls the actions of the body.

The spinal cord carries sensory information (touch, pain, body position) to the brain and relays motor information from the brain to muscles, organs, and glands. Interneurons in the brain and spinal cord integrate incoming and outgoing signals so the body responds appropriately.

The spinal cord is also responsible for reflexes, which are immediate reactions to stimuli, like when you pull your hand away from a hot stove. They're fast because they bypass the brain. Your hand moves before your brain registers "hot."

The Peripheral Nervous System

The peripheral nervous system (PNS) is a network of nerves running from the CNS throughout the body. The PNS connects to the CNS via the cranial nerves that attach to the brain and the spinal nerves that branch out from the spinal cord. The PNS is divided into two systems:

- o Somatic: Carries signals to and from the skeletal muscles, usually in response to stimulus, and this is under your conscious control.
- o Autonomic: This is involuntary and regulates the internal environment of the body by operating the heart muscles and other organs and glands.

The autonomic nervous system is divided into the parasympathetic and sympathetic branches. The parasympathetic branch helps the body do things to gain and conserve energy. They sympathetic branch prepares the body for intense action. These two systems constantly interact to maintain optimal body function in any circumstance.

Sensing

All of our senses are perceptions constructed by the brain based on information from sensory receptors. You don't really see with your eyes It's all what your brain is interpreting. Humans have five sensory receptors:

1. Pain

2. Thermoreceptors: temperature

3. Mechanoreceptors: touch, pressure, motion, and sound

4. Chemoreceptors: chemicals

5. Electromagnetic receptors: light and electricity

How is it all coordinated?

Neurons are arranged in local circuits that interconnect and communicate. These form systems and the brain is the system of systems where all mental functions are implemented. So function is both localized and interconnected with the entire body.

Why did Gage die from seizures?

It's not clear how Gage's initial brain injury would cause seizures, but maybe subsequent jostling caused further damage. Seizures indicate the brain is misfiring and can't get back in balance. This information goes to the muscles and they contract, so the body twitches and moves uncontrollably. This would give off so much energy as heat that the body temperature would begin to drop. In response, the brain would shut down circulation in the hands and feet but if cooling continued, circulation to the major organs would stop and Gage, in this case, died.

13.3 Define these Key Terms

neurons

nerves

Central Nervous System

Spinal cord

Interneurons

Reflexes

Peripheral Nervous System

Autonomic

Somatic

Sympathetic

Parasympathetic

13.3 Questions for Review

13.3.1 Draw a diagram of the NCS and PNS and indicate the general parts of each.

13.3.2 What is a reflex (example?) and why is it a faster response than normal?

13.3.3 What is the difference between the functions of the somatic and autonomic nervous systems?

13.3.4 What does the sympathetic nervous system do? The parasympathetic?

13.3.5 List the functions of the 5 sensory receptors in humans.

13.3.6 Explain the statement "You see with your brain, not your eyes."

13.4 How Does Cell Communication Lead to Nervous System Function?

Learning Objective 13.4:

Describe the process of a neuron firing, and predict the impact of changes to neuron function.

13.4 Summary

Neurons are highly specialized signaling cells that exchange signals with other cells. They are long and skinny with projections or dendrites branching off of one end. The other end, the axon, reaches to the next neuron. The end of the axon has short branches ending in a terminal bud that transmits information. The gap between the axon of one neuron and the dendrite of the next is called a synapse.

Messages travel along the axon as an electrical wave called an action potential. At the axon terminal, the electrical signal is converted to a chemical signal that is picked up by the receiving cells and converted back to an electrical signal. The signal travels like a very fast bucket brigade.

How Neurons Talk to Each Other

When a neuron is at rest, the ionic charge from the salt in the body fluids is different on one side of the cell membrane than on the other. The outside of the cell is positively charged. The cell membrane keeps proteins and other large molecules inside the cell, and these are usually negatively charged.

Now let's say there is a sudden, loud noise. The sensory neurons send information to the CNS and the sympathetic branch of the autonomic part of the PNS signals danger. The adrenal gland releases a neurotransmitter called adrenaline. Neurotransmitters bind to specific receptors on the neuron membrane to initiate a nerve impulse or action potential. In this case, adrenaline triggers the opening of the ion channels in the cell membrane, allowing positive ions to rush into the cell.

This flow of positive ions triggers the membrane channels in the axon to open, allowing those positive ions in, and so on. Once the nerve impulse travels all the way to the end of the axon, vesicles, or membrane-enclosed bubbles containing adrenaline, fuse with the cell membrane, and adrenaline squirts into the synaptic space. It then binds to the membrane of the next neuron and the whole process repeats until the signal reaches its destination or effector. In the case of adrenaline, the effectors would include heart muscle and the response would be an increased heart rate. In humans and other vertebrates, the axons are encased in an insulating, fatty, myelin sheath, which speeds transmission. A nervous impulse travels down an axon as fast as 100 meters per second or 200 miles per hour.

What Neurons Say to Each Other

There are at least 25 known neurotransmitters. Each is associated with particular characteristics and often they work together such that a neuron is bathed in a mixture. Different mixtures produce different responses and some magnify or dampen the response. We are constantly bombarded with stimuli and the nervous system must sort out what to pay attention to and what can be ignored.

13.4 Define these Key Terms

dendrite

axon

terminal bud

synapse

action potential

neurotransmitter

13.4 Questions for Review

13.4.1 When the neuron is at rest, draw in where you would find positive and negative charges.

13.4.2 Describe the "bucket brigade" that is a nerve transmission. What might trigger it, and how does it move through the cell and then on to the next cell?

13.5 What If the Nervous System Is Injured or Diseased?

Learning Objective 13.5:

Detail how injuries and disease can impact nervous system function.

13.5 Summary

Injury to any part of the nervous system disrupts communication with potential physical, emotional, and/or cognitive issues.

Injury

A localized injury to the brain, like Phineas Gage's, results in loss of the specific function of that region. Sometimes a brain injury can cause a seizure disorder like epilepsy. A violent blow to the head can result in a concussion, or brain bruise, in which the electrical activity in the brain is temporarily disordered. Spinal cord injury can have different outcomes depending on the location. If the damage is severe, there will be loss of sensations and possibly paralysis below the site of the injury.

Infection

Even with modern antibiotics, people suffer from infections of the nervous system. Meningitis is an inflammation of the meninges. It is caused by a bacterial or fungal infection, is contagious and often fatal. Encephalitis is a rare inflammation of the brain, usually due to a viral infection. This is often carried by mosquitoes and has a 33% mortality in humans.

Neurodegenerative Diseases

Neurodegenerative diseases are numerous and fairly common. There are more than 40 syndromes in which neural communication breaks down because neurons of the CNS deteriorate and die. Here are a few:

- Alzheimer's Disease (AD): More than 5 million Americans suffer from AD. It usually appears after age 65 and starts with memory loss, degenerating until the person cannot perform daily activities. AD patients have an abnormal protein, beta amyloid, gumming up their axons. There is no prevention and no cure.
- Parkinson's Disease (PD): More than a million Americans have PD. It usually begins between 50 and 60 years of age. Patients experience a gradual loss of motor control due to lack of the inhibitory neurotransmitter dopamine. Without dopamine, excitatory signals from the motor regions of the brain trigger shaking and tremors. There is no cure or prevention.
- Multiple Sclerosis (MS): At least 400,000 Americans suffer from MS. It is an autoimmune disease meaning the body's own immune system attacks itself, and is a major cause of disability in young adults. The myelin sheath is progressively destroyed, slowing nerve impulses and triggering weakness, stiffness, speech slurring, and difficulty walking
- Amyotrophic Lateral Sclerosis (ALS) or Lou Gehrig's disease: ALS afflicts 30,000 Americans. It attacks motor neurons causing muscles to shrink and atrophy. Eventually the neurons for breathing and swallowing are affected. ALS is always fatal but sensing and thinking remain completely functional until death.

13.5 Define these Key Terms

concussion

meningitis

encephalitis

Alzheimer's disease

Parkinson's disease

Multiple Sclerosis

Amyotrophic Lateral Sclerosis

13.5 Questions for Review

13.5.1 Describe three types of neural injury and note a possible consequence of each one.

13.5.2 Describe two types of brain infections.

13.5.3 Fill in the following table of neurodegenerative diseases, in order from most to least common:

Disease	What does the disease do?

13.6 Can Mental Health Be Separated from Physical Health?

Learning Objective 13.6:

Evaluate the role of drugs on behavior, mental health, and overall human health.

13.6 Summary

Because thought, behavior, emotions, and feelings all derive from brain function, it doesn't make much sense to separate physical from mental health. Depending on the specific cause and symptoms, some mental health problems can be treated successfully. There are two commonly diagnosed mental health issues.

1. Attention Deficit Hyperactivity Disorder (ADHD): Six percent of school-aged boys and 2% of girls in the United States take Ritalin to treat diagnosed ADHD. Symptoms include forgetfulness, impulsiveness, distractibility, fidgeting, and impatience. In some cases, these symptoms are due to low dopamine levels.

2. Depression: This affects 20 million adults in the United States, two-thirds of them women. Symptoms include sadness, loss of energy, changes in weight and sleep patterns, and a general loss of interest. In severe cases, the person is unable to function and in an extreme case may commit suicide. It's caused by an imbalance of neurotransmitters, particularly low serotonin. Medications raise serotonin levels, and this is often combined with psychotherapy.

Drug Use and Abuse

Many substances that make us feel energized, happy, relaxed, or sleepy do so by inserting themselves into some aspect of nervous system communication. Some are highly addictive and dangerous. Whatever the drug, if the person can't control themselves they have a mental health problem. This is a list of some common recreational drugs:

- o Alcohol: A depressant; works by increasing the effects of GABA, an inhibitory neurotransmitter.
- o Caffeine: A stimulant; counters the effects of inhibitory neurotransmitters and increases the metabolic activity of all cells.
- o Nicotine: A stimulant; activates acetylcholine receptors and causes the cerebral cortex to produce acetylcholine, epinephrine, and norepinephrine.
- o Marijuana: THC in marijuana binds to receptors in the brain that play a role in pain, depression, appetite, and memory, stimulating some of these functions and inhibiting others.
- o Cocaine: A stimulant; increases release of norepinephrine and dopamine, can lead to symptoms similar to schizophrenia.
- o Ecstasy: A stimulant and hallucinogen; causes release of serotonin which can produce panic, seizures, high blood pressure and heart rates, dramatically increased body temperature, organ failure, and death.
- o Lysergic acid diethylamide (LSD): Hallucinogen that increase the normal response to serotonin.
- o Amphetamines: Stimulants that work by mimicking the effects of norepinephrine and inhibiting the enzyme that normally breaks down norepinephrine.
- o Tranquilizers: Depressants that activate GABA receptors, an inhibitory neurotransmitter (Valium and Xanax).
- o Opiates: Codeine, morphine, and heroin, which bind to endorphin receptors reducing pain and producing euphoria.

13.6 Define these Key Terms

Attention-Deficit Hyperactivity Disorder

Depression

Psychotherapy

13.6 Questions for Review

13.6.1 What is ADHD and why is its diagnosis somewhat controversial?

13.6.2 What is depression and how is it treated?

13.6.3 What distinguishes drug use from drug abuse and a mental health problem?

13.6.4 Fill in the following table describing various types of "recreational" drugs:

Drug	Stimulant or depressant?	Effects

13.7 Can Nervous System Injuries and Illnesses Be Cured?

Learning Objective 13.7:

Assess the progress and potential of treatments for nervous system injury and disease.

13.7 Summary

Physical damage to the brain or spinal cord cannot be repaired. Neurons do not regenerate. Scientists are investigating this though.

Drugs

There are medications to address symptoms of various neurodegenerative diseases, and drugs being developed to address the disease mechanisms and correct the problem.

Surgery

Surgical implants of normal dopamine-releasing tissues have helped alleviate the symptoms of PD. Physicians have tried transplanting the Schwann cells that make the myelin sheath to help those with MS. Trials have been done using gene therapy to treat AD.

Regeneration and Stem Cells

Many people think that using stem cells to regenerate and replace damaged neural tissue has the best chance of curing neurological disorders and repairing CNS injuries.

Minimizing Damage

Techniques are being developed to minimize spinal cord damage resulting from an accident. Drugs that reduce inflammation and realigning the spine may both prevent an injury from becoming more serious. Chilling the patient for a few days to slow metabolism and minimize swelling helps. Rehabilitation is always necessary.

Chilling also helps in brain injuries, as does opening the skull to relieve pressure.

Prevention, using helmets and seatbelts, remains the best defense.

13.7 Define these Key Terms

13.7 Questions for Review

13.7.1 Describe methods being investigated for treatment of spinal cord injuries.

13.7.2 Why does chilling a patient help prevent damage in spinal cord and brain injuries?

13. Biology in Perspective

As scientists learn more about how the nervous system works, our ability to diagnose and treat neurological disorders has advanced. Onset of some diseases can be predicted or prevented, and we're better able to deal with CNS injuries. Scientists can also manipulate our nervous systems. This raises questions: should we medicate eccentric people? Should we try to identify potentially harmful behavioral tendencies in brain activity? To what extent is our humanness a product of brain and nervous system function? Do we have free will?

Question to consider:

As our understanding of neurobiology advances, What are some things that scientists/doctors are better able to do? List some examples.

13. Scientist Spotlight: Linda B. Buck

How did Linda Buck's research impact what we know about odors and our sense of smell?

At Columbia University and then Harvard, Dr. Linda Buck discovered more than 1000 genes that control the production of chemoreceptors that bind odorant molecules, and she showed that these are found in the upper part of the nose. She and colleagues also found that each cell in this area has only one type of odor receptor and each receptor can only detect a limited number of odors. Once an odor is detected, a signal is sent to the olfactory bulb. Information from several olfactory receptors is combined to produce what we experience as a distinct odor. Buck won a Nobel Prize for this work in 1988.

Question to consider:

Describe how we "smell" particular odors.

13. Technology Connection: Measuring Brain Activity

What might brain imaging be able to tell us about the likelihood for developing diseases like alcoholism or schizophrenia?

Until the 1990s, the only way to determine what parts of a brain did what was to stimulate or disable those areas in animal brains and observe the response, or to study people who had damage in that particular area. In the 1990s, the technique of functional magnetic resonance imaging (fMRI) was developed; it measures blood flow in the brain. The brain requires a lot of energy, and active regions receive extra blood. Active areas "light up" in fMRI. So far, this has allowed scientists to learn a lot about the neural circuits involved in sensations, movement, and language. This information aids in planning brain surgery and addressing chronic pain. At some point we may be able to use this technique to predict the likelihood of developing diseases like alcoholism or schizophrenia, or tendencies for behaviors like aggression or pedophilia.

Question to consider:

What does fMRI allow scientists to do and how does it work?

13. Life Application: Concussion

How can you get a concussion without hitting your head?

A concussion is a brain bruise caused by your brain slamming into the inside of your skull. This can be caused by falls, accidents, violence, sports, or simply sudden shaking or deceleration, like in a car accident. You don't actually have to hit your head. In some cases the collision between your brain and skull results in bleeding and tearing of the delicate nerve fibers. In extreme cases, brain swelling can produce widespread neurological damage and death.

If you receive a blow to the head, symptoms to look for are a dazed feeling, a headache, nausea, amnesia, confusion, dizziness, or slurred speech. It's essential to minimize further brain trauma until the injury has healed, so an athlete should never go back into a game after a particularly hard hit. Multiple concussions over a lifetime results in cumulative neurological damage that cannot be fixed.

Question to consider:

What, exactly, is a concussion? What can cause one?

13. How do we know: The Pleasure Centers of the Brain

Is it possible to experience happiness with the flip of a switch?

Scientists planned to study a portion of the brain in a rat that, when electrically stimulated, caused the rat to run away from a particular area of a maze. Instead of running away, however, the rats became quite attracted to the area. The scientists realized they could direct the rats to go anywhere by stimulating this particular portion of the brain. The stimulation evidently made the rats feel "good." The scientists then trained the rats to stimulate themselves by moving a lever. It turned out that rats would rather move that lever than eat, drink, mate, or take care of their offspring. They would even endure an electrical shock to get to the lever.

It turns out that there are several regions deep in the brain that form a pleasure or reward circuit. Humans react to stimulation in these areas much like the rats.

Question to consider:

Explain how the "pleasure" circuit in the brain was found.

Student Study Guide

Chapter 14: Infectious Disease and the Immune System

How Are Invaders Repelled, Evaded, or Killed?

Our bodies contain an entire ecosystem of tiny microorganisms—mostly bacteria and fungi—that help to keep us healthy, and most of the time it works. Sometimes there is an invasion that disrupts the system, though. Our immune system recognizes the invaders and works to eliminate the problem. Sometimes it is effective, and sometimes it is not, for various reasons. So how does it work?

14. Case Study: The Summer of Sorrow

Why did so many Irish immigrants die in the 1840s?

14. Case Study Summary

In 1847, a fungus destroyed the entire potato crop in Ireland, the primary source of food for the tenant farmers. In a country of 8 million people, 2 million died of starvation, and more than a million immigrated to the United States and Canada. The immigrant ships were crowded and dirty, and many people did not survive the journey or were sick when they landed.

These immigrants were blamed for the spread of cholera, which was highly contagious and often fatal. Grosse Isle, outside Quebec in Canada, was used to isolate incoming immigrants. During the summer of 1847, 398 ships sailed into Grosse Isle. More than 5000 people died during these trips, and more than 5000 were buried in mass graves after dying once they'd reached the island. More than 9000 people eventually died from a cholera outbreak that summer. Yet not everyone died. What happened?

Question for Review

Explain why immigrants were blamed for the cholera outbreak on Grosse Isle.

14.1 What Causes Cholera?

Learning Objective 14.1

Describe the history and biology of cholera.

14.1   Summary

Cholera is an infection of the intestines that can kill a person in 24 hours as a result of dehydration due to diarrhea and vomiting. It can be treated by rehydrating patients, but that was not known for a long time, and even now the disease commonly kills people, particularly in underdeveloped nations, or following natural disasters like the earthquake in Haiti in 2010. How does cholera spread?

John Snow's Explanation for How Cholera Spreads

There was a cholera outbreak in London in 1848 and 1849. John Snow was a doctor there, and in treating patients, he noticed that the disease often occurred in people who got their drinking water from the same source. In 1854 there was another outbreak, and Snow carefully mapped out who got sick and who didn't. He identified the source of the disease as sewage contamination at the Broad Street water pump. He concluded that individuals were catching the disease from drinking water contaminated with the feces of sick people. Officials didn't believe him, but they dismantled the pump and the cholera outbreak stopped.

Robert Koch's Explanation for What Causes Cholera

During the mid- to late 1800s the bacteria responsible for diseases such as anthrax, tuberculosis, and diphtheria were discovered by a group of scientists, including Robert Koch. He was sent to India in 1883 to figure out what bacteria caused cholera. He examined the feces of cholera patients and found *Vibrio cholera.*

How Cholera Affects the Body

Individuals with a severe case of cholera can lose gallons of fluid within a day. The salts or electrolytes necessary for muscle function and water balance also flood out of the body. They go into shock due to lack of blood circulation. People in shock have dangerously low blood pressure, feel faint and dizzy, and are pale. If rehydration does not occur, 50–60% of individuals with cholera experience multiple organ failures. But some exposed people never get sick, and some of the sick recover on their own.

How *Vibrio* Bacteria Survive and Spread Cholera

Trillions of harmless bacteria live on and within our bodies. Potentially dangerous bacteria don't harm us most of the time. But when conditions are right (or wrong), some bacteria can be deadly. In the case of *Vibrio*, this is what it has to do to cause trouble:

- Get into the body
- Survive the trip through the stomach and into the intestines
- Evade the body's defenses
- Attach to cells and take up residence in the intestines
- Reproduce to make more bacteria
- Trigger the river of diarrhea that will deliver the bacteria to the next host

Acid in the stomach is the first line of defense, but if you swallow enough bacteria at once, that may not be sufficient. If the *Vibrio* make it into the intestine, they stick to the cell membranes there, and once attached, they make more sticky molecules in order to stay put. They then secrete a toxin that binds to another molecule on the intestinal cell membrane, enters the cell, and binds to the G protein, which disrupts the system controlling water balance. Chloride rushes out into the intestine, sodium and other electrolytes follow, and water then follows the salt via osmosis. Experiments with people drinking different levels of the toxin have demonstrated that this is what causes the diarrhea, and then the *Vibrio* effectively surf out to the next host. In crowded, dirty conditions, it's easy to ingest enough bacteria to become ill.

14.1 Define these Key Terms

cholera

Vibrio

toxin

14.1 Questions for Review

14.1.1 How did John Snow figure out the source for cholera in the London outbreaks? How do we know he was correct?

14.1.2 What did Robert Koch contribute to our understanding of cholera?

14.1.3 Draw a flow chart illustrating the progression of a cholera infection in an individual. For each of your boxes in the chart, explain what happens.

14.1.4 Why do cholera outbreaks tend to occur in crowded, dirty conditions?

14.1.5 How does cholera actually kill a person?

14.2 What Invaders Do We Face?

Learning Objective 14.2:

> *Describe the array of potential pathogens humans are exposed to every day.*

14.2 Summary

Bacteria are relatively simple prokaryotic cells, meaning they lack membrane-bound organelles. Thousands of species live on and in us all the time. We're also surrounded by microorganisms. Mostly, these cause us no harm. Some help us digest, and some keep bad ones from invading. Bad ones include the bacteria that cause tuberculosis, strep throat, gonorrhea and syphilis, "flesh-eating" staph infections, and food poisoning. Bacteria reproduce quickly: one cell could produce 1 billion cells in about 10 hours. And many pathogenic bacteria produce toxins.

Protozoa and other Eukaryotes

Protozoa are single-celled eukaryotes, which means they are cells like our own, with membrane-bound organelles. These cause giardiasis, amoebic dysentery, sleeping sickness, and malaria. Fungi cause athletes foot, jock itch, ringworm, and yeast infections. From contaminated meat or fish, we can get tapeworms, round worms, blood flukes, or guinea worms.

Viruses and Prions

Viruses are not exactly alive because they can't reproduce unless they are in a host cell. They contain DNA or RNA, and they get into a cell and take over its metabolism to make virus proteins, nucleic acids, and more viruses. These are responsible for AIDS, chicken pox, and the common cold.

Prions are infectious proteins. Normally prion-like particles are found in the brain, but when they change to an abnormal form, they become pathogenic, forming clumps in the brain that damage cell structures. These produce degenerative brain diseases like mad cow disease, scrapie (sheep), and Creutzfeldt-Jacob disease (humans). Luckily, these are rare diseases. Prions are the only pathogens invisible to our immune system.

14.2 Define these Key Terms

prokaryotes

eukaryotes

protozoa

viruses

prions

14.2 Questions for Review

14.2.1   What is the difference between a prokaryote and a eukaryote? Give an example of each.

14.2.2 List the eukaryotic organisms that can be pathogenic.

14.2.3 How is a virus not a living thing? How does it "live" then?

14.2.4 What is a prion and where are they found?

14.2.4 Below is a list of human diseases. Indicate whether each is caused by bacteria, protozoa, fungi, worm, a virus, or a prion.

Mad cow, yeast infection, Creutzfeldt-Jakob, strep throat, tuberculosis, ringworm, chicken pox, AIDS, malaria, staph infection, giardiasis, common cold, blood flukes, tapeworms, syphilis

14.3 How Does the Immune System Protect Us?

Learning Objective 14.3:

> *Explain how the human immune system works, and predict the effect of the removal of a portion of the system.*

14.3 Summary

Not everyone exposed to cholera gets sick. Maybe they just avoided swallowing the bacteria. Maybe their stomachs were extra acidic. People with AB blood are resistant to cholera while those with O are particularly susceptible. It's not clear why. Carriers of the gene associated with *cystic fibrosis* are quite resistant, perhaps because the chloride transporter *Vibrio* hijacks in the intestine doesn't work right. But everyone has an immune system at work trying to prevent infection.

The First Line of Defense: Don't Let the Pathogens Get In

First of all, your general state of well-being influences how vulnerable you are to infection. Nutrition, age, climate, fatigue, and socioeconomic status all matter.

Non-specific barriers help keep pathogens out. Skin presents a barrier because it's thick and mostly dry, and as we sweat and slough off skin cells, pathogens can be swept away. Mucous membranes in the digestive tract, reproductive tract, respiratory tract, and urinary system trap pathogens. Saliva contains antimicrobial enzymes. Acid in both the stomach and vagina destroy bacteria. Cells in the respiratory system sweep mucus and any trapped microbes up into the throat to be swallowed. Flowing urine keeps the urinary tract free of bacteria. Tears contain antimicrobial chemicals.

The Second Line of Defense: Don't Let the Pathogens Make a Home

If pathogens make it through the first line of defense, there are three more non-specific internal defenses:

> 1. Defensive cells: Macrophages locate and kill microbes by engulfing and eating them. Natural killer (NK) cells kill virus-invaded cells by penetrating their outer membranes, causing the cells to burst

> 2. Defensive proteins: Cells infected by viruses produce interferons, chemical signals that enlist the help of non-infected cells to fight the virus. Complement proteins coat the surface of the bacteria, making it easier for the macrophages to eat it. These also poke holes in the pathogens so they burst.

> 3. Inflammatory response: Macrophages send out chemicals to increase blood flow and make the blood vessels leaky in an infected area. This causes painful swelling but limits the infection to the local region so the specific immunity has time to work.

The Third Line of Defense: Know Your Enemy

Rehydrating a cholera patient usually allows them to recover, but it's not the rehydration itself that cures the disease. Rehydration gives the body time to muster its own specific immune response, and that is what offers the cure. There are two types of specific immune response; both involve white blood cells called lymphocytes.

1. Antibody-Mediated Immunity: B Cells

The body produces B cells continuously in bone marrow. Receptors on the membranes of these cells recognize specific foreign molecules or antigens on invading cells. Each B cell recognizes only one type of antigen, and it binds to that one so the cell can't reproduce. The B cells also make and secrete antibodies that also bind to the specific antigen. The antibodies circulate through the body, binding to the invading pathogen and complement proteins join in to poke holes in the cells, killing them. A small number of B cells become memory cells that then remain in the body for years, ready to attack that same pathogen if it invades again. The B cells specialize in dealing with bacterial infections.

2. Cell-Mediated Immunity: T Cells

Bone marrow also produces T lymphocytes or T cells that then mature in the thymus. There are two types. Cytotoxic T cells specialize in attacking and killing cells infected by viruses, cancer cells, fungi and worms. Helper T cells assist. First, macrophages engulf and clump together the foreigners. T cells that can bind to that specific antigen will then reproduce more T cells that can recognize the invader. Helper T cells detect foreign invaders, enhance B cell responses, alert other T cells that there is an infection, and increase the response of cytotoxic T cells. The T cells also produce memory cells to aid in any future infection.

Sometimes the immune system works too well, as is the case when transplanted organs or tissues are rejected because they are "foreign" to the immune system.

When the Immune System Fails

The immune system doesn't always work. Sometimes it attacks its own cells, resulting in autoimmune diseases like rheumatoid arthritis, multiple sclerosis, or type I diabetes. Poor health or depression can dampen the immune response. Immunosuppressive diseases like AIDS can disable the immune response. In that case, the helper T cells are greatly reduced. Finally, pathogens change so fast that B and T cells "remembering" one virus from a past infection may not recognize the newest version. This happens with the influenza (flu) virus, and also malaria.

14.3 Define these Key Terms

macrophages

natural killer cells

interferons

complement proteins

inflammation

lymphocytes

B cells

Antigens

Antibodies

Cytotoxic T cells

Helper T cells

14.3 Questions for Review

14.3.1 Draw a diagram or flow chart of the path of a potential pathogen through the defensive systems of the body. Assume that the pathogen gets through the defenses until it is finally stopped by specific immunity. Do separate paths for a virus and a bacterial infection.

14.3.2 Explain why organ rejection occurs when organs are transplanted.

14.3.3 Describe three cases in which the immune system doesn't work? What happens in each?

14.4 How Can We Harness the Immune System?

Learning Objective 14.4:

Assess the strengths and weaknesses of natural immunity, vaccination, and passive immunity.

14.4 Summary

Natural Immunity

This is a consequence of the memory cells produced by our specific immune system. If we've already had and recovered from an infection, the next time we're exposed, we might not be infected at all, or the infection will be mild.

Vaccination

The strategy of vaccinating people against pathogens uses the natural immunity produced by the specific immune system. The vaccine is a weakened form of the pathogen or some component or product of it. Memory cells are produced that recognize this pathogen if we are exposed to it later in life. Unfortunately, some pathogens change their surface proteins over time and so would be unrecognizable by memory cells. Malaria and the flu virus are like this. In some cases, like cholera, a vaccine offers only partial protection. In that case, prevention by watching what you eat or drink is the best option.

Passive Immunity

Babies and children have not yet been exposed to many diseases so they are particularly vulnerable, but they receive passive immunity from their mothers. Antibodies circulating in the blood of the mother cross the placenta to enter the fetal bloodstream. And the first breast milk also contains antibodies. This immunity is temporary, so we also vaccinate babies.

14.4 Define these Key Terms

natural immunity

vaccination

passive immunity

14.4 Questions for Review

14.4.1 Describe how natural immunity works.

14.4.2 How do vaccines work?

14.4.3 Why do vaccines not work for every pathogen?

14.4.4 Babies have underdeveloped immune systems when they are born. What helps them avoid infection at these early stages?

14.5 What Can Help If Our Immune System Fails?

Learning Objective 14.5:

Assess the use and describe the production of antivenom, antibodies, and antibiotics.

14.5 Summary

Luckily, we have a few strategies for treating ailments when the immune system fails.

Antivenom

If you are bitten by a venomous snake, you have no antibodies to recognize the venom, and it acts too fast for your immune system to respond. You will need antivenom. This is made by injecting goats or horses with increasing amounts of venom so that their immune systems make antibodies. These are then extracted from the blood and concentrated to serve as a treatment for a venomous bite.

Antibodies

A person can sometimes be injected with antibodies isolated from someone else in order to boost their immune system. Monoclonal antibodies (MAbs), which target only diseased cells, are another source.

Antibiotics

People are, of course, sometimes overcome by infections. In these cases, we can use antibiotics. Microorganisms compete with each other, so they produce chemicals to inhibit or kill each other. We can sometimes use these to treat infections effectively.

14.5 Define these Key Terms

antivenom

antibiotics

14.5 Questions for Review

14.5.1 Where does antivenom come from and how does it work?

14.5.2 Why might antibodies from someone else help in an infection?

14.5.3 Where do antibiotics come from?

14.6 How Can We Prevent the Spread of Infectious Disease?

Learning Objective 14.6:

Predict the impact of personal choices on your health and the health of those in your community.

14.6 Summary

As it turns out, we have considerable control over our health, and this includes avoiding infection.

Wellness

The reason why cholera took hold of the Irish immigrants so easily was, in part, because they were already malnourished and fatigued, and then they were crowded together. When you are physically weakened, your immune system is weakened, too.

Hygiene

You can't get cholera if you don't swallow *Vibrio*. Frequent hand washing helps for all infectious diseases. Don't cough or sneeze without covering your nose and mouth. Don't sneeze into your hand, though. Facemasks help if it's an infectious respiratory pathogen. Condom use and not sharing needles or syringes decreases the likelihood of running into contaminated body fluids.

Clean Water and Sewers

Countries with well-developed water and sewage treatment systems eliminate the biggest source of diarrheal disease pathogens. Cholera is rare in the developed world.

Proper Antibiotic Use

If an antibiotic is needed to cure an infection, the infected person does not have the natural immunity that would come from his or her own immune system. If a population is infected by cholera and they're all given antibiotics, the next time cholera shows up, they're all vulnerable again. In addition, pathogens can develop resistance to antibiotics when they are not used correctly.

Vaccines

If you get a vaccine for a particular disease, you won't get the disease. But you also protect the population. If everyone in a population is vaccinated, everyone is safer because even if the vaccine is not 100% effective, the pathogen can't gain a foothold. If only 50% of the population is vaccinated, the pathogen can stay around, and some people can't tolerate vaccines because of other health issues. If the pathogen can stay around because a lot of people choose not to be vaccinated, everyone is more vulnerable.

14.6 Questions for Review

14.6.1 Compare the present-day U.S. society and the situation for the Irish immigrants. What made one of these populations more vulnerable to cholera than the other?

14.6.2 Describe how you can avoid acquiring an infection other than by vaccine.

14.6.3 Why is it better to allow people to recover from a disease without resorting to antibiotics if at all possible?

14.6.4. You should understand by now how vaccines work to protect the vaccinated individual from infection, but how does getting as many people as possible vaccinated help protect everyone in the population?

14. Biology in Perspective

Infectious diseases are still a problem, and avoiding them requires individual and societal efforts. Our immune systems are quite effective, but personal behavior matters too. It is also important that everyone in society has access to clean water, food, shelter, waste treatment, and preventative medical care.

Question to consider:

Why does the general condition of people in a society (the socioeconomic well being) matter when it comes to preventing infectious disease?

14. Scientist Spotlight: Anthony S. Fauci

Why was Dr. Anthony Fauci accused of being a murderer by HIV/AIDS protestors?

Dr. Anthony Fauci was director of the National Institute of Allergy and Infectious Disease (NIAID) and in charge of the fight against HIV/AIDS in the United States in 1988. Activists were clamoring for access to experimental drugs and were protesting Fauci. He invited them in to hear their concerns, and this led to the participation of activists in many phases of planning. It also led to the idea that gravely ill people can get access to experimental drugs.

Dr. Fauci is still director of NIAID, where he oversees both basic and applied research in the prevention, diagnosis, and treatment of infectious diseases like HIV/AIDS, sexually transmitted diseases, influenza, tuberculosis, and malaria. He has developed therapies to successfully treat formerly fatal inflammatory and immune-related diseases. He has also helped reveal how the immune system works and how immunosuppressive drugs dial down the immune response. He has been at the forefront of basic research on how HIV destroys the body's defenses.

Question to consider:

Why was Dr. Anthony Fauci accused of being a murderer by HIV/AIDS protestors, and what did he do about it?

14. Technology Connection: Monoclonal Antibodies: Magic Bullets?

Are there drugs that target only diseased cells, while leaving healthy cells unharmed?

In 1975, Georges Kohler and Cesar Milstein wondered if it might be possible to isolate and grow a population of the B cells to make a desired antibody. The problem was that B cells have a finite lifespan. Continuous repetition and re-growth would be required to maintain the supply of antibodies. To solve this problem, they set about making immortal B cells. They injected an antigen into mice to elicit an immune response and then they collected the B cells. These were fused with a blood cancer cell called a myeloma and this resulted in immortal cells that each produced one type of antibody. They then grew these cells. Because they came from one clone of B cells, they're called monoclonal antibodies. MAbs are used to detect specific molecules. Many are already in use. Their advantage is that they target only diseased cells, avoiding side effects for healthy cells.

Question to consider:

What do monoclonal antibodies do?

14. Life Application: Organ and Tissue Transplantation

How many people's lives can be helped or saved through the donation of one person's tissues and organs?

While it is certainly a difficult decision under awful circumstances, one person who has succumbed to some traumatic injury rendering them "brain dead" can help or save the lives of as many as 50 other people. In the example in this section, the family of a boy injured in a car accident donated his liver, lungs, both kidneys, heart, and corneas.

The immune system will reject this transplanted tissue as foreign and attack it, but there are strategies to overcome this. First, the donor and recipient tissues are matched, which means they share 75% of their "identity proteins." Second, the patient receives immunosuppressive drugs, but these also make the person more vulnerable to infection.

These techniques save lives, but there are more lives to be saved than there are organs to go around. Scientists are currently working on promising techniques for growing tissue from the patients themselves.

Question to consider:

You are a biologist and you are trying to explain to your brother why organ donation is a good idea. Make your case.

14. How Do We Know?: The Immune Response is Two-Fisted

What role did a loan of some chickens play in the discovery of the major components of the immune system?

Sometimes important discoveries are accidents.

In the 1950s, Bruce Glick was a graduate student studying a structure called the bursa of Fabricius (BF) in chickens. He wanted to understand what it did, so he removed it from hatchlings and monitored the birds.

A fellow graduate student, Timothy Chang, was studying the immune response of chickens to the bacterial pathogen *Salmonella*. He needed to inject the birds with *Salmonella* antigens so he could collect antibodies from the chickens. He asked if he could use Glick's chickens.

None of the injected chickens produced any antibodies at all. It turned out that the BF is where B cells begin developing in embryos. The non-BF birds still exhibited the cell-mediated immune response, and it was thought that the cells necessary for this originated in the thymus. As it turned out, when the thymus was removed from newly hatched chicks, cell-mediated immunity was lost. If the birds lacked both a thymus and a BF, they had no specific immunity.

Question to consider:

How did two graduate students accidently discover where B-cell production occurs?

Student Study Guide

Chapter 15: Nutrition, Activity, and Wellness

How Can We Live a Healthy Lifestyle?

All of your body systems work together, so health problems don't usually start with a problem with just one system. It's all interconnected. Wellness, or good health, relies on the balanced functioning of all our systems. To a great extent, your lifestyle can determine how balanced and healthy you are.

15. Case Study: Dying for the Perfect Body

What drove a successful ballet dancer and a teenage baseball player to make unhealthy decisions that would cost them their lives?

15. Case Study Summary

Heidi Guenther, the Dancer

Heidi Guenther was a gifted dancer from childhood and started ballet school at age 13. But by age 18, she no longer had the lithe body shape of a child. At age 19, Heidi joined the Boston Ballet Company and was living her dream. At that time she was 5'4" tall and weighed 110 pounds. She injured her foot, and due to less physical activity while it healed, she gained 5 pounds, which she was told to lose. A doctor prescribed diet pills and she lost 10 pounds. She started smoking to control her weight, dropped to 93 pounds, and still thought she was overweight. Other dancers thought she was purging and using laxatives. On her family's annual summer trip to Disneyland, Heidi collapsed in the back seat of the car. Her muscles tensed, her lips turned blue, and resuscitation efforts failed. She died of cardiac arrest at age 22.

Taylor Hooton, the Baseball Player

Taylor Hooton thought he had the opposite problem from Heather's. He was an outstanding high school baseball player, and he was 6'2" and 180 pounds, but he thought he should be bigger. He started using steroids to build up muscle quickly. This resulted in uncharacteristic violent rages at home, and his family confronted him with their suspicion that he was using steroids. He promised to stop. Suddenly stopping steroids use usually triggers depression, so Taylor's doctor prescribed an antidepressant. Upon returning from a trip, Taylor's parents found him preparing to sell computer equipment to finance the steroid habit he had not managed to quit, and they grounded him. In response, he hung himself.

These two people suffered from body image problems, something very common. But there are healthy ways to slim down or bulk up.

Questions to Review

Fill in this table comparing the circumstances of Heather and Taylor:

	Body image problem	How did he/she deal with the problem?	How did their respective "solutions" kill them?
Heather			
Taylor			

15.1 How Does What You Eat Influence Your Well-Being?

Learning Objective 15.1

Identify and describe the components of a healthy diet.

15.1 Summary

Food Quantity and Quality

We eat food together to be sociable and celebrate, and it is the source of energy that drives everything the body does. Food molecules contain energy in the chemical bonds that hold atoms together. Our bodies extract this energy and convert it into a form we can use. We need enough food to supply the energy the body needs. No single food provides everything we need, so we also should eat a variety: grains, vegetables and fruits, proteins, and small amounts of unsaturated fats (fats from vegetables).

Water

The simplest and most essential component of a healthy diet is water. It is necessary for all cellular activities, so a person cannot survive for more than a few days without it. We need to take in about 3 liters of water per day to replace what is lost in sweat, urine, and feces. About half that typically comes from food. Water is necessary to maintain normal blood pressure, to eliminate wastes, to maintain body temperature, and to move nutrients and oxygen through the blood and into tissues and cells.

Carbohydrates, Proteins, and Fats

Most of our energy comes in three forms:

1. Carbohydrates

These are sugars found in bread, cereal, rice, pasta, fruits, and vegetables. The more complex carbohydrates are generally healthier than the simple sugars because they are digested more slowly and therefore release energy gradually.

2. Proteins

Proteins are made of amino acids. There are 20 types of amino acids used to build all our body proteins. We can synthesize 12 of them, and the other 8 are "essential," meaning we need to eat them regularly. They cannot be stored. Fish, meat, eggs, grains, nuts, beans, and dairy products are all proteins. In addition to providing energy, proteins build structures and regulate chemical reactions in the body.

3. Fats

Fat or lipids are molecules composed of fatty acids. They yield twice the energy of carbohydrates or proteins. The average person stores enough body fat to supply 4–5 weeks of energy. Our bodies can synthesize most of the fatty acids it requires, but a few "essential" fatty acids must be eaten. Fats come from meats, dairy products, nuts, and vegetable oils. Fats also insulate the body to maintain heat, act as "shock absorbers," and serve as building blocks for hormones and other signal molecules.

In addition to these three, we also require a number of micronutrients, so called because we don't need much:

- o Vitamins: These act as co-enzymes, molecules that assist enzymes in the regulation of specific chemical pathways.
- o Minerals: These are inorganic molecules (they do not contain carbon). They are necessary for bone and tooth development and health, muscle contraction and relaxation, nerve impulses, and fluid balance in the body.
- o Fiber: Fiber is composed of complex carbohydrates that we can't fully digest. It helps lower cholesterol and low-density lipoproteins (LDL or "bad" lipids) in the blood. It keeps things moving in the digestive tract and so lowers the risk of cancers of the large intestine and colon.

15.1 Define these Key Terms

carbohydrates

fats

proteins

vitamins

minerals

fiber

micronutrients

amino acids

15.1 Questions for Review

15.1.1 Fill in the following table describing the items we need to consume in our diets.

Dietary need	What is it made of?	What does the body need it for?	What foods supply it?
Carbohydrates			
Proteins			
Fats			
Vitamins			
Minerals			
Fiber			

15.1.2 What does it mean when something in our diet is termed "essential?"

15.2 How Does the Body Extract Nutrients from Food?

Learning Objective 15.2:

Trace the path that food travels from the mouth to the bloodstream, and identify the key places where nutrients are extracted.

15.2 Summary

The digestive system is a 30-foot continuous tube from mouth to anus. Its various regions break down ingested food, absorb the food molecules into the cells, and extract energy from the chemical bonds in those molecules. This chapter describes the structures and process.

From Mouth to Stomach

Lips hold the food in the mouth. Teeth break down food into smaller pieces by chewing. The tongue lets you know how the food tastes. Usually, spoiled food smells bad enough that you know not to ingest it at all, but sometimes the tongue lets you know. Salivary glands secrete saliva that both lubricates the food and contains enzymes to begin the breakdown of carbohydrates into simple sugars. The tongue shapes the food into a ball that's easy to swallow into the pharynx at the back of the throat. The pharynx leads to the esophagus that leads to the stomach. The pharynx also leads to the windpipe or trachea to the lungs. The epiglottis normally closes the trachea when we swallow. When it fails we choke.

The Stomach and Small Intestine

Food moves through the esophagus into the stomach by waves of muscle contractions called peristalsis. The stomach can hold 3 liters of material and a ring of muscle called a sphincter controls the opening between the esophagus and stomach. Heartburn happens when stomach contents back up into the esophagus.

Cells in the stomach lining secrete acid that digests proteins and kills most bacteria that might be swallowed. They also secrete enzymes that break down proteins and mucus to protect the stomach from acid damage. The semi-liquid stomach contents, called chyme, then move to the small intestine.

The small intestine is about 20 feet long and 1 inch in diameter. Most chemical digestion and then absorption of nutrients into the blood stream occurs here. The key to these activities is the very large surface area: the small intestine is the size of a tennis court. A sphincter controls movement of chyme into the small intestine.

The Pancreas and Liver

The pancreas connects to the small intestine near where the stomach connects, and it secretes materials that neutralize the acidic chyme coming from the stomach. It also secretes enzymes that digest carbohydrates, proteins, and fats into sugars, amino acids, and fatty acids, respectively.

The liver produces bile that dissolves fats. Bile is stored in the gall bladder. The liver also stores some vitamins and extra glucose, the most commonly used simple sugar the body uses for energy. It can provide an energy boost to the body. It can also convert excess carbohydrates and proteins into fat for long-term storage as adipose tissue.

From the Intestine to the Bloodstream

Once the food molecules have been broken down into simple subunits, these nutrients are transported into the cells lining the small intestine, then into the fluid space between the cells, and then into the bloodstream. These are then transported to the cells and tissues that need them.

Indigestible material moves into the large intestine or colon. This is 3 to 6 feet long and 3 inches in diameter; it absorbs water, salts, and some vitamins. The last 6 inches are the rectum, where feces are stored prior to elimination.

15.2 Define these Key Terms

salivary glands

pharynx

esophagus

epiglottis

peristalsis

chyme

pancreas

bile

gall bladder

glycogen

liver

adipose tissue

colon

rectum

15.2 Questions for Review

15.2.1 Draw a diagram of the digestive system from mouth to anus. For each region, list the name and what goes on there.

15.2.2 What is digestion?

15.2.3 What are the risks of repeated purging of food by vomiting, as the dancer Heidi was likely doing?

15.3 How Do Cells Extract Energy from Food?

Learning Objective 15.3:

Explain the process and chemistry of cell respiration.

15.3 Summary

Food is broken down to sugars, amino acids, and fats, and the energy in these molecules needs to be transformed so that it can be used by the body. Cell respiration does this.

Cell Respiration

Cell respiration involves the slow "burning" of energy-rich molecules until the energy is used up and just carbon dioxide and water are left. Enzymes in the cell gradually extract the energy and store it in other molecules the body can use.

The Chemistry of Cell Respiration

There are three steps to the process of the energy extraction:

1. Glycolysis: In the cytoplasm of cells, the sugar glucose is broken down to pyruvic acid. This process releases two molecules of ATP, the energy currency of the cell for practically all chemical actions that require an input of energy.

2. Kreb's cycle: The Kreb's cycle occurs inside the mitochondria of the cell and produces GTP and NADH—both energy currencies like ATP—from the pyruvic acid. The NADH is then used in step 3 below.

3. Electron transport chain: Also in the mitochondria, the NADH produced above is used to activate an enzyme that produces more ATP.

15.3 Define these Key Terms

cell respiration

glycolysis

Kreb's cycle

ATP

NADH

Electron transport chain

15.3 Questions for Review

15.3.1 Draw a diagram of a cell and indicate where each of the three steps in cellular respiration occur. For each step, what is broken down and what is produced?

15.3.2. What does a cell do with ATP?

15.4 What Can a Person Do to Maintain a Healthy Weight?

Learning Objective 15.4:

Describe how a person's basal metabolic rate can be used to maintain a healthy weight.

15.4 Summary

Calories are the units of energy used to describe food. If you eat more calories than you use up, you will gain weight as your body stores the leftover energy as fat tissue. If you use up more energy than you eat, you will lose weight.

Basal Metabolic Rate

Even when you are at rest, your body is using energy to support itself. This "at rest" is your basal metabolic rate. This typically ranges from 1300 to 1800 calories per day, depending on sex, weight, age, genetics, and activity level.

You can determine how many calories you should consume per day by taking the weight you want to maintain and multiplying by 11. Then add the number of calories you expend during exercise. The more you exercise, the more you need.

Average Dimensions

What is defined as "attractive" depends on culture. For example, the average American woman is 5'4" tall and weighs 166 pounds. The average female runway model is 5'10" and 100–120 pounds. Obviously we need to distinguish between cultural and biological definitions of being over- or underweight.

The Body/Mass Index

The BMI is a helpful tool for determining whether weight is in the healthy range, based on one's height and weight. A healthy BMI is generally 20-25. This method isn't perfect; it doesn't take into account gender, frame size, or muscle mass, but it's a good indicator. At 5'4," when Heidi died at 93 pounds, her BMI was 16, dangerously underweight.

15.4 Define these Key Terms

Calories

basal metabolic rate

BMI

15.4 Questions for Review

15.4.1 What is a Calorie a measure of?

15.4.2. Describe how you gain and lose weight, in terms of Calories.

15.4.3 How is your basal metabolic rate different from your active metabolic rate? Which one is higher? Why?

15.4.4 What is the purpose of calculating one's BMI?

15.5 What If Weight Becomes an Obsession?

Learning Objective 15.5:

Assess the various dangers associated with eating disorders that commonly afflict individuals.

15.5 Summary

Sixty percent of American adults are obese. That is unhealthy in itself, increasing the risk of heart disease, stroke, high blood pressure, joint problems, and certain cancers. Just as unhealthy, though, are eating disorders where people try to quickly slim down or bulk up through some kind of drastic actions.

Eating Disorders

In theory, losing weight simply requires that you burn more calories than you take in, but this is difficult in modern society where we are surrounded by food. Millions of individuals in the United States suffer from an unhealthy relationship with food and an obsession with body weight and image. This can result in an eating disorder, where a person eats too little or too much to be healthy. There are several types.

- o Anorexia nervosa: Victims of this are always unsatisfied with their bodies and starve themselves, sometimes also exercising obsessively, in order to lose more weight. Peak of onset is puberty to young adult.
- o Bulimia nervosa: Individuals with this disorder binge eat, then feel guilty and purge by vomiting or using laxatives.
- o Binge-eating disorder: A binge-eater will gorge on high-Calorie food, often in secret, in effort to fulfill perceived emptiness. They often feel embarrassed but do not purge. This is associated with depression and low self-esteem.

These disorders were once considered mostly "female problems," but men suffer from them as well.

Loss of minerals as a result of anorexia leads to kidney damage, muscle cramps, irregular heart rhythms, and even cardiac arrest. Depletion of body fat to produce hormones needed for the menstrual cycle can result in sterility. Insufficient estrogen can result in osteoporosis or thinning bones. Bulimia puts a person at risk for all of that plus dental and gum problems from all the stomach acid, stomach rupture from repeated vomiting, and potentially fatal dehydration.

These diseases are especially prevalent among athletes and other performers for whom thinness matters. Recovery is possible with mental health counseling and support.

Steroid Abuse

This happens when a person desires to be larger and stronger. Steroids are drugs that mimic normal sex hormones like testosterone and make muscles bigger. They also affect other body systems, however, causing shrinkage of the testicles and reduced sperm count, increased body hair and smaller breasts in women, kidney and liver damage, heart disease, acne, and psychological problems in men and women.

15.5 Define these Key Terms

anorexia

bulimia

binge eating

steroids

15.5 Questions for Review

15.5.1 Explain why a dancer or fashion model might be especially prone to an eating disorder.

15.5.2. Fill in the table:

Eating disorder	Associated behavior	Health consequences

15.5.3 List the side-effects of steroid abuse.

15.6 How Do Muscles and Bones Keep Us Active?

Learning Objective 15.6:

Describe the components of the motor system and the ways muscles and bones work together to enable us to move.

15.6 Summary

Body movement is governed by the actions of the skeleton and the muscles. The skeleton supports the body, protects soft parts, and stores minerals, especially calcium. Bones produce blood cells. We have 206 bones. The skull and spine are the axial skeleton and the hips, shoulders, and limbs make up the appendicular skeleton.

The hinges where bones connect, held together by ligaments (bone to bone), allow movement when worked by muscles. The hips and shoulders are ball and socket joints, allowing three-dimensional range of motion. The elbows and knees are hinge joints. All joints are cushioned by a flexible tissue called cartilage. Muscles attach to bones with tendons (muscle to bone). Skeletal muscles contract and bones move. We have more than 700 muscles, and they account for about 40% of our weight.

We also have cardiac muscle, located in the heart. The regular heartbeat requires a huge amount of ATP, so cardiac muscle cells have more mitochondria than any other body cells. Cardiac muscle is not under our conscious control, and neither is smooth muscle that is build into blood vessels, the bladder, intestines, and uterus.

Moving Muscles

Muscles are made of parallel filaments. Each filament is made of muscle fibers in which each fiber is a cell controlled by one neuron. Within each cell are long, parallel myofibrils composed of as many as 100,000 contraction units called sacromeres, attached end to end. The sacromeres are made of the proteins actin and myosin, anchored to z-disks.

At the signal to contract, the thin actin filaments slide over the myosin filaments, pulling the z-disks closer to each other. This process requires ATP. Skeletal muscles are arranged in antagonistic pairs (biceps and triceps, for example), so when one contracts (sacromeres shorten), the other relaxes (sacromeres lengthen).

15.6 Define these Key Terms

skeleton

ligament

cartilage

tendon

skeletal muscle

cardiac muscle

smooth muscle

sacromeres

actin

myosin

15.6 Question for Review

15.6.1 Describe the sequence of events which allow you to raise your arm, beginning with the brain deciding to do that.

15.7 How Can You Keep in Shape?

Learning Objective 15.7:

> *Discuss the safety and efficacy of various methods for getting and staying in shape.*

15.7 Summary

Taylor Hooton put on 30 pounds. of muscle in 6 months. He could not have done this so quickly by exercise alone, but since there is no safe level of steroid use, exercise is by far the better option. How does it work?

Exercise

Muscles are composed of slow-twitch and fast-twitch fibers. Fast-twitch fibers contract and relax quickly. They fatigue quickly. Slow-twitch fibers are slower to contract and relax but have more endurance. Sprinters use more fast-twitch fibers and long-distance runners use more slow-twitch.

To build up muscle strength, weight lifters train with weights set at 80–90% of the maximum weight they can lift once. This type of training causes muscle cells to make more actin and myosin, and more fibrils mean more strength.

Training with lower weights but more replicates will increase muscle size but not muscle strength.

To increase both strength and size, both types of training are required. Strength training improves bone health and endurance too. In addition, aerobic exercise strengthens the heart.

Supplements

There are a few drugs or nutritional supplements that athletes may try, other than steroids. They have side effects and their manufacture and dosages are not highly controlled so they are risky and don't work as well as exercise, good nutrition, and sleep.

- o Diruetics: These alter the fluid balance in the body, resulting in high volumes of urine, reducing weight but potentially causing dehydration, muscle cramps, fatigue, and irregular heartbeat.
- o Creatine: This is normally found in the body and helps muscles make more ATP, delaying muscle fatigue. Excess amounts though, can cause muscle and stomach cramps, nausea, diarrhea, and weight gain due to water retention.
- o Vanadium: This supplement is supposed to increase muscle mass, but there's no evidence that it does, and it causes diarrhea and stains the tongue green.

- Boron: This is a nutritional supplement—for plants.
- Stimulants: These can reduce fatigue and suppress appetite, improve alertness, and sometimes make a person more aggressive. They can also cause irritability, nervousness, insomnia, increased blood pressure, heart abnormalities, heart attack, and hallucinations.

15.7 Define these Key Terms

slow-twitch muscle fibers

fast-twitch muscle fibers

diuretics

creatine

vanadium

stimulants

15.7 Questions for Review

15.7.1 What are fast-twitch and slow-twitch muscle fibers used for? How are they different?

15.7.2 If you wanted to increase the amount of weight you could lift, describe the training you would do?

15.7.3 Fill in the table:

Supplement	Why you would take it?	Possible side effects
Diuretic		
Creatine		
Vanadium		
Stimulant		

15.8 What Is the Difference Between Looking Good and Being Healthy?

Learning Objective 15.8

Distinguish between what it means to look good versus what it means to be healthy and why the two are not always synonymous.

15.8 Summary

"Real" people don't look like our images of fashion models, actors, and athletes. An obsession with weight and body image to look like these can have tragic consequences. Most people know what is required to be healthy: a balance, varied diet emphasizing whole grains, fruits, and vegetables and minimizing highly processed, fatty foods and alcohol, regular exercise, adequate sleep, and managing stress.

Physical Activity

Regular physical activity prevents or improves coronary heart disease, heart attack, stroke, diabetes, high blood pressure, hip fractures, and obesity. It benefits muscle strength, bone health, and joint structure and function. It also improves mental health, reduces the risk of depression, improves performance in school, and bolsters self-esteem.

Sleep, Laughter, and Managing Stress

Scientists and health care professionals recommend 7–9 hours of sleep per night. One of the principal benefits is to reduce stress. Stress increases blood pressure, the risk of stroke and heart attack, and causes inflammation. It also makes it harder to sleep, which further increases the stress hormones. Consequently, sleep deprivation increases the risk of heart disease, diabetes, and cancer. Also, when darkness falls, a molecule called melatonin increases in the body, and this seems to suppress tumor growth. Staying up with the lights on reduces melatonin production. Sleep also bolsters memory, improves alertness, reduces the risk of depression, and decreases the risk of being overweight. Without enough sleep, the balance of hormones controlling appetite causes us to feel hungry when we don't really need to eat.

Laughter reduces the level of stress hormones in the body and increases beneficial brain chemicals like endorphins, which reduce pain and elevate mood. It also enhances the effectiveness of immune system cells.

15.8 Define these Key Terms

stress

melatonin

15.8 Questions for Review

15.8.1. Describe what you need to do to maintain good health.

15.8.2 List as many consequences of sleep deprivation as you can remember, then go back and read and include what you forgot.

15.8.3 List benefits of sleep.

15.8.4 Why is laughter helpful in maintaining health?

15. Biology in Perspective

The recipe for wellness is easy to understand: eat a varied, balanced diet; exercise or be physically active at least 30-60 minutes most days; sleep no fewer than 7 hours a night; don't drink excessively; don't use tobacco; don't use drugs for recreation; don't buy into media images of how we should all look; and laugh a lot. There is no other pathway to good health.

Question to consider:

What do you need to do (or not do) to maintain your health?

15. Scientist Spotlight: Linus Pauling

How did Linus Pauling's life's work both transform our understanding of chemistry and contribute to world peace?

Linus Pauling studied chemical engineering at what is now Oregon State University, and at Cal Tech he studied chemistry. Eventually he won two Nobel Prizes, in chemistry and for peace.

Pauling's career had four phases. First, his discoveries transformed our understanding of chemistry. Second, as a political activist, he championed nuclear disarmament and world peace. Third, he established the field of molecular medicine by showing that sickle cell disease is due to a faulty gene. This interest in chemistry and health led to his fourth phase, which he called orthomolecular medicine. This had to do with maintaining the optimum balance of chemicals in the body to maintain health.

Question to consider:

Linus Pauling is about as famous as a scientist can get, yet his advocacy of taking high doses of vitamin C to maintain optimum health is not something most people do. Why not?

15. Technology Connection: "Guaranteed" Six-Pack Abs

All kinds of claims are made in advertisements for fitness machines, but studies have shown that you can't "spot reduce." If you need to lose weight somewhere, you need to lose it everywhere. Trying to tone individual groups of muscles, like abs, can be done, but it makes more sense to do overall muscle conditioning, aerobic workouts, and eat well for overall appearance and health.

Question to consider:

Find an advertisement for some kind of fitness machine and compare what it says to what would actually make sense if you were trying to do what it claims to do.

15. Life Application: Drunkorexia

What risky weekend college pastime is leading many women to endanger their health?

Drunkorexia refers to the phenomenon of college-aged women skipping meals in order to save calories for alcohol consumption. Women may also drink shots of straight alcohol rather than mix it in order to avoid the extra calories in the mixers. Alcohol normally affects women more quickly than men due to their physical differences, and if one couples that with drinking on an empty stomach, one can get really drunk really quickly. In these situations blackouts are more common, as is vomiting and possible dehydration. More than 30% of college students meet the criteria for a diagnosis of alcohol abuse and roughly 25% of anorexics and bulimics have drug or alcohol issues.

Question to consider:

Why is avoiding eating in order to save calories to drink more alcohol a really bad idea?

15. How Do We Know?: Stress Weakens the Immune Response

How can stress make you sick?

It has been shown experimentally that stress increases your susceptibility to disease. In one study, people were intentionally exposed to a cold virus or just a saline solution, and the people undergoing more stress in their lives were much more likely to actually get a cold as a result of the exposure. Studies suggest that this is because stress suppresses the immune system response.

Question to consider:

Explain how we know that stress increases the likelihood of getting sick.

Student Study Guide

Chapter 16: Ecology

How Do We Benefit from a Functional Ecosystem?

Ecology is the study of interactions between living things and their environment. Species compete, exploit, cooperate, and adapt, and energy and nutrients cycle throughout the ecosystem. We all depend on the work of ecosystems for our survival.

16. Case Study: The Near Extinction of the Kirkland's Warbler

What drove the Kirkland's warbler to near extinction in the 1970s?

16. Case Study Summary

Kirkland's warblers are rare and nest only in a few sites in northern Michigan. In 1973 a census detected only 200 males, for a total population of about 400 birds. Why so rare? There are a few hypotheses:

- o Their food source is rare.
- o Something is eating them.
- o Some disease is killing them.
- o Human development is crowding them out.
- o Suitable habitat and nesting sites are rare.

So which is it? We need to know some Kirkland's warbler ecology. They build their nests on the ground, under the lower branches of young jack pine trees. Young jack pine trees are short and spaced apart, so shrubby plants grow well between them, providing protection and food for the warblers. While jack pines are not rare, young jack pines are. Once the trees get tall, they shade out the plants on the ground, including new jack pines. They also drop their lower branches. This leaves warbler's nests exposed.

The only way jack pines reproduce is if the pine cones burn and the older trees die back a bit from forest fires, which used to be relatively common. The sandy soil in which jack pines live is also perfect habitat for human homes, though, and the soil is also good for agriculture. And fire is not good for either of those, so we've done our best to suppress forest fires.

Finally, cowbirds, which follow cattle and bison around to eat insects they stir up, and so do not make their own nests, lay their eggs the nests of other birds. Agriculture has opened up the jack pine forests such that the cowbirds have invaded, and the bigger, aggressive cowbird babies often get most of the food so the baby warblers starve.

So human development, limited nesting habitat, and nest parasitism seem to largely account for the decline in Kirkland's warblers. There is now a program in place in Michigan to conduct controlled forest burns, trap cowbirds, and educate people about these birds. As of 2010, there were 1747 breeding pairs.

Questions for Review

16.1 List some reasons any species of wild animal might be undergoing a population decline?

16.2 What type of habitat do Kirkland's warblers require?

16.3 Why is the habitat required by the Kirkland's warblers rare now?

16.4 What role do cowbirds play in this decline of the warblers?

16.1 How Do Species Adapt to Their Habitat?

Learning Objective 16.1

Provide examples of the ways species adapt to the constraints of their habitats.

16.1   Summary

Many physical factors interact to determine a habitat suitable for a particular species: temperature range, water availability, food availability, shelter. Species adapt to habitat both physiologically and behaviorally.

Adapting to Physical Conditions

A desert is a difficult place to survive because maintaining water balance is a challenge. Kangaroo rats are an example of how an animal can do this. Like other animals, they lose water through urine, feces, and evaporation, but they lose much less than normal through urine and feces because of their extremely efficient kidneys. Their feces are also dry because their colon recovers almost all the water. To minimize evaporation from respiratory surfaces, the rats are nocturnal and spend their days in a burrow, which is humidified by their exhalations. Food stored in the burrow then absorbs some of that water, and they are able to re-ingest it. Like other animals, they also produce water internally via the metabolic process. So these little rats never drink, but they take in about 40% more water than they lose, and so they can live in the desert.

Adapting to Limited Shelter

Many animals have specific requirements for shelter and for reproduction. The yellow-bellied marmot, for example, lives high in the Rocky Mountains. They have dens in the cracks and crevices of the rocks. There aren't enough of these for every marmot to have its own, so one male claims and defends a den for 2–5 females. If the sex ratio is 50:50, in a polygynous ("many females") mating system like this, not every male gets to mate.

White-fronted bee eaters are birds that live in central Africa and dig cavities in riverbanks for their nests. There is often not enough space for every bird to nest, so young bee-eaters frequently stay with their parents for an extra year, helping to raise their younger siblings.

16.1 Define these Key Terms

ecology

polygynous

16.1 Questions for Review

16.1.1 Draw a diagram illustrating the inputs and outputs of water for a kangaroo rat. What special adaptations to the desert does it have and how do they work?

16.1.2 How do marmots adapt to a lack of shelters in their habitats?

16.1.3 How do bee-eaters adapt to a lack of shelters in their habitats?

16.2 Why Do Species Compete?

Learning Objective 16.2:

Summarize how species can coexist without one outcompeting the other.

16.2 Summary

Competition between species occurs when two or more try to use the same resource. Both species are harmed (or limited). If one benefits and the other is harmed, it is exploitation. If both benefit, it is mutualism (or cooperation). Benefit and harm are measured in population size (abundance) and distribution (where it can live).

The Competitive Exclusion Principle

Competition happens when resources are limited. If both species are harmed by the competition, it would be best to avoid it. How does this work? One competitor can drive the other to extinction, at least in that one area, or one or both species can change in some way so they no longer need the same resource.

The scientist F.G. Gause experimented with two microorganisms and found that *Paramecium aurelia* always drove *Paramecium caudatum* to extinction when they were grown together. Apart, they both did fine. He generalized that when any species compete, the best competitor will always win and exclude the other from the habitat: the competitive exclusion principle. This holds up well in lab conditions where options are limited, but in nature, conditions are often variable enough in one location to allow species to coexist. One might feed at night and one during the day, for example. A whole suite of resources promote the growth, survival and reproduction of a species. This is called its niche: biological factors like food and nesting sites, and physical factors like temperature and water. According to the competitive exclusion principle, no two species can occupy the exact same niche.

Instances When Competitors May Coexist

The Galapagos Islands are home to 14 species of finch, made famous by Charles Darwin. Finches typically eat seeds determined by their beak size. The medium ground finch and the small ground finch have about the same sized beaks but they live on different islands, so they don't compete for the same size seeds. On the island of Santa Cruz, however, they live together, and these same two species have different sized beaks here, so they avoid competing. A change in a physical or behavioral characteristic which minimizes competition like this is called character displacement.

16.2 Define these Key Terms

competition

exploitation

mutualism

competitive exclusion principle

niche

character displacement

16.2 Questions for Review

16.2.1 Indicate whether you think each of the following examples is exploitation, competition, or mutualism. For each type of organism indicate a + if it gains, a – if it loses. Try to fill in a few examples of your own.

Organisms	Plus or minus for each?	What's the relationship?
Wolves and elk	/	
Clownfish (Nemo!) and anemones	/	
Zebras and Wildebeasts	/	
Ladybugs and aphids	/	
Cowbirds and Kirkland's warblers	/	

16.2.2 The last section discusses kangaroo rats. Describe their niche.

16.2.3 You might agree that polar bears and grizzly bears are rather similar animals. Apply the competitive exclusion principle to them. If they both started out in North America, how might that have driven them to be different in the ways that they are? (Think about their adaptations.)

16.2.4 Explain what character displacement is and how it reduces competition.

16.3 How Do Species Exploit Each Other?

Learning Objective 16.3:

Describe the ways that species may exploit one another.

16.3 Summary

Exploitation occurs when one species benefits as a result of an interaction, at the expense of the other. Here are some examples:

- o Predation: A predator captures prey to eat
- o Browsing and grazing: Organisms consume plants without killing them.
- o Parasitism: One organism lives off of another.
- o Disease: One organism uses another as a host.

These result in population cycles, and evolutionary "arms races."

Population Cycles

A predator–prey system is a good example of this. In North America, Canadian lynx eat snowshoe hares, and because both have been trapped for a century, we have long term population data. The hare population rises, then the lynx rises until they overexploit the hares. The hare population crashes, then the lynx population crashes, and so it goes, about every 10 years. As usual though, the more we know, the more complicated it gets. The hare and the shrubbery they eat have their own population cycle, and lynx eat other prey, too.

An Evolutionary Arms Race

As the exploiter develops adaptations to better exploit, the exploited species develops adaptations to avoid exploitation. They coevolve. For example, many plants produce secondary compounds, chemicals that are poisonous or unpalatable to grazers. The grazers may develop resistance. Monarch butterflies and milkweeds are an example of this.

In another example of coevolution, the European cuckoo lays its eggs in the nests of other birds. Some of these other birds recognize that those eggs don't match theirs (discrimination) and they kick them out of the nest. To get around this, individual female cuckoos specialize in parasitizing particular species nests. Their eggs mimic the eggs of their chosen host. This egg mimicry is a response to discrimination.

An evolutionary response takes time. Kirkland's warblers currently fall victim to parasitism from cowbirds. Cowbirds have only lately invaded their territory though. Maybe there just hasn't been time to adapt.

16.3 Define these Key Terms

predation

grazing

parasitism

16.3 Questions for Review

16.3.1 Describe an example of each of the ways organisms can exploit each other.

16.3.2 Draw a graph illustrating the population cycles of the lynx and snowshoe hare.

16.3.3. There are two examples of coevolutionary arms races described in this chapter. Think about animals they exploit and see if you can describe another example.

16.3.4 Why haven't the Kirkland warblers adapted to the nest parasitism from the cowbirds?

16.4 When Can Species Cooperate?

Learning Objective 16.4:

Identify circumstances of cooperation between species.

16.4 Summary

In cooperation or mutualism, both species benefit. It's mutual exploitation. Animals pollinate plants and disperse their seeds. In return the plant provides pollen, nectar, fruit, or shelter. Honeybees pollinating plants are a great example of this. They pollinate many types of plants. Some pollinators specialize, like each fig wasp species specializing on one type of fig. It lives in and feeds off only that fig. In this case, the mutualism is obligatory. One can't survive without the other.

Plant–pollinator relationships can break down due to cheating. Robber bees get nectar without performing pollination. Some orchids have flowers that resemble female bees or wasps, and may even give off pheromones to attract males. The males show up, attempt to copulate, and get covered with pollen. When they move to the next flower to try again, they transfer that pollen.

16.4 Define this Key Term

mutualism

16.4 Questions for Review

16.4.1 Explain how honeybees offer an example of mutual exploitation or cooperation?

16.4.2 What do the bees and the plants each get out of this relationship?

16.4.3. Explain an example of obligatory mutualism.

16.4.4 Explain how a plant or animal can "cheat" in these interactions.

16.5 How Do Ecological Interactions Affect Us?

Learning Objective 16.5:

Summarize the beneficial and detrimental ecological interactions that we have with other species in the ecosystem.

16.5 Summary

Humans are part of the ecosystem on Earth, too.

We Compete with Other Species

Humans are intelligent, and we work together to compete effectively for the resources we need. In many cases, our competition drives other species to extinction. But when we change habitat, opportunities for some species open up. Farm fields provide good habitat for certain insects. Mice, molds, and seed-eating insects do well in grain silos. In fact, many species we consider pests are those that are good competitors with us. The others just disappear.

Annually, humans spend billions of dollars on pesticides and herbicides in the United States alone, not counting research, oversight, health costs, and cleanup. In spite of this, food loss to pests ranges from 20% to 50% of the potential yield. This is especially a problem for developing countries. And the use of pesticides and antibiotics may lead to stronger pests in the long run. Biological control, where natural predators and parasites are used to fight pests, is promising, but introducing new organisms to an environment can have unintended consequences.

Early in our history, our species, *Homo sapiens*, and *Homo neanderthalis* co-existed in Europe for 10,000–15,000 years, but eventually *H. sapiens* was the only species left. We were, evidently, more flexible in our diet, more tolerant of a changing climate, and possibly better hunters.

We Develop Useful Products and Ideas as a Result of Exploitative Interactions

Early in our history we were involved in predator–prey interactions and may well have driven many prey species extinct. More recently, we hunted for food, fur, and sport. Now most of our food is domesticated.

We've been good at taking advantage of the adaptations found in nature. For example, the cardiac glycosides produced by milkweeds have been used to treat heart conditions. Caffeine in coffee plants deters competition from other plants, and herbivores don't generally like coffee beans. We like that caffeine, though.

Most of the spices we use are produced by plants as chemical defenses. They may repel insects or kill bacteria. Hallucinogenic drugs from plants and fungi have been used for thousands of years. Some have medicinal properties as well.

We Capitalize on Mutualisms

Fruit is a great example of mutualism. Fruits contain the seeds of a plant, and protect them, and aid in dispersal. Animals like to eat the fruits and then carry off the seeds, depositing them away from the parent plant in a nice pile of fertilizer. We've taken advantage of this long-standing mutualism, and we have bred larger and tastier fruits.

16.5 Define this Key Term

biological control

16.5 Questions for Review

16.5.1 Draw a flow chart diagramming what happens when humans alter a natural habitat. Who wins and who loses?

16.5.2 What kinds of species do we most actively compete with in this modern age?

16.5.3 Describe an example of humans outcompeting some other organism.

16.5.4 List a few examples of how humans take advantage of chemical defenses produced by plants.

16.6 What Does a Functioning Ecosystem Do?

Learning Objective 16.6:

Describe the movement of energy and nutrients through an ecosystem.

16.6 Summary

Exploitation Interactions Distribute Energy and Nutrients

Energy in an ecosystem comes primarily from the sun. Plants capture sunlight in the form of energy-rich chemicals, sugars, and starches. They use the energy to drive their own metabolism, and when they are eaten, the energy gets transferred. Predation distributes energy and nutrients through the ecosystem. When organisms die, their bodies return nutrients to the soil for the plants.

The energy flows through a food web made up of the following trophic levels:

- Primary producers: the photosynthetic organisms that capture the sun's energy
- Primary consumers: the herbivores
- Secondary consumers: carnivores (can continue to tertiary, etc.)
- Decomposers: organisms that release and recycle the energy from dead bodies

Energy flow through an ecosystem is not efficient. Numerically, if we start with 100%, on average, only 10% of that gets passed along to the primary consumers. And only 10% of that gets to the herbivores, and so on. About 90% is lost as heat and indigestible material in each level. This means higher trophic levels can support very few individuals. There are many more antelope than there are lions, for example. To illustrate this, the food web is often drawn as an ecological pyramid.

Ecosystems Recycle Material on a Global Scale

The matter on Earth is stored in four reservoirs:

- Biosphere: all the life on the planet
- Atmosphere: gaseous layer surrounding the Earth
- Hydrosphere: water on Earth
- Lithosphere: rock

Biogeochemical cycles move matter between these four reservoirs. Note that they include "bio-," so living things play an important part in this.

The Carbon Cycle: Photosynthesis and respiration drive the movement of carbon through an ecosystem. Photosynthesis captures carbon dioxide and converts it to carbohydrates. As organisms break down carbon compounds they release carbon dioxide back to the atmosphere

The Oxygen Cycle: Photosynthesis adds oxygen to the atmosphere and respiration removes it. Burning also consumes oxygen, but because oxygen is 21% of the atmosphere while carbon dioxide is only 0.04%, fires don't have an appreciable impact.

The Nitrogen Cycle: Nitrogen cycling is complicated because it involves a wider variety of species working on a sort of biochemical assembly line. Nitrogen makes up 79% of the atmosphere, but no plant or animal can use it directly, and yet we all need it to build proteins and DNA. Soil bacteria make it available to us in a two-step process. Some soil bacteria convert atmospheric nitrogen gas to chemicals like ammonia. Other soil bacteria convert that to nitrates that plants can absorb. Urine and decomposition return it to the soil, and other soil bacteria convert it back to nitrogen gas and it goes back to the atmosphere.

16.6 Define these Key Terms

food web

trophic levels

primary producers

primary consumers

secondary consumers

decomposers

ecological pyramid

16.6 Questions for Review

16.6.1 Draw a diagram of a ecological pyramid, filling in the layers. For each layer, provide an example of an organism that would be at that level.

16.6.2 Explain why there can't be as many lions as there are zebras in the African savannah ecosystem.

16.6.3 Draw a flow chart illustrating the carbon cycle.

16.6.4 Add arrows to show the oxygen cycle along with your carbon cycle.

16.6.5 Draw a flow chart illustrating the nitrogen cycle

16.6.6 Describe the role the decomposers play in the biogeochemical cycles.

16.7 What Benefits Do We Get from a Functioning Ecosystem?

Learning Objective 16.7:

Enumerate the benefits we get from a functioning ecosystem.

16.7 Summary

Ecosystems are interesting, but if we want to truly preserve ecosystems, we probably need to consider the economic benefits of a functioning ecosystem, as well as the costs of one that is not functioning.

Ecosystem Services

Ecosystem services are naturally occurring processes that support agriculture, technology, culture, and more. There are four categories:

- o Provisioning services: These provide us with food, fresh water, natural fibers, fuel, and biochemical compounds, and genes.
- o Regulating services: These regulate and stabilize the ecosystem, making it more reliable. These include regulating pests and disease, controlling runoff and erosion, pollination, water and air purification, and waste decomposition.
- o Cultural services: These relate to quality of life issues and are uniquely human. They include the recreational, aesthetic, and inspirational value of natural areas, the cultural, spiritual, and religious connections people have to nature, and the joy of learning about nature and the opportunities it provides for scientific discovery.
- o Supporting services: These are what supports the whole system and include the capture of solar energy, biogeochemical cycling, and the production of soil and oxygen.

We already assign economic value to the provisioning services: they provide items we can use or sell. Historically, we have not placed economic value on the others but we are learning that if one of these is disrupted, there is a cost that we really need to take into account in our development projects.

Disruptions to Ecosystem Services

All terrestrial life depends on soil and our agriculture, forestry, and development practices have shifted the balance between soil loss and soil erosion such that we lose soil globally, annually. Development projects need to take in to account the cost of replacing soil.

In exploiting other species, we can disrupt the food web. If we overhunt a predator, the population of herbivores it normally eats will expand and overeat the plants. The ecosystem can collapse and be unable to provide services.

We can disrupt biogeochemical cycles, too. Burning forests and fossil fuels releases more stored carbon dioxide to the atmosphere than photosynthetic organisms can absorb. This additional carbon dioxide causes global climate change. In addition, we apply nitrogen to our agricultural fields in the form of fertilizer, rather than relying on soil bacteria. This enhances crop production but the excess nitrogen leaches into water and promotes algal growth.

Biodiversity

All of the ecological processes discussed depend on a diverse set of species in the ecosystem. Reducing biodiversity is one of the most common results of development.

16.7 Define this Key Term

biodiversity

16.7 Questions for Review

16.7.1 Provide an example of each of the four ecosystem services.

16.7.2 Explain why it should be important to take into account the economic value of ecosystem services when we develop in some way which might impact them.

16.7.3 Describe four ways humans can disrupt ecosystem services.

16. Biology in Perspective

We benefit from the interactions among species. Many of their adaptations are useful to us, providing food or medicine. Our survival depends on the benefits the ecosystem provides. Maintaining these services intact is an economic imperative, as well as an ethical and cultural need.

Question to consider:

Outline the reasons we should worry about maintaining intact ecosystems in our world?

16. Scientist Spotlight: Robert Helmer MacArthur

How did Robert MacArthur bring mathematical and experimental rigor to the study of ecology?

Dr. Robert Helmer MacArthur was initially interested in natural history but in school he studied mathematics, along with zoology. His main contribution was to bring mathematical and experimental rigor to ecology. He developed mathematical models that showed the relationship between species numbers and habitat size that could then be tested. His research provided a theoretical foundation for ecology.

Question to consider:

What did Dr. MacArthur provide to the field of ecology that was different than what had come before?

16. Technology Connection: Biological Control of Schistosomiasis

How could crayfish prevent parasitic flatworms from infecting humans?

Schistosomes are parasitic flatworms that burrow into human skin, find a blood vessel, and travel throughout the circulatory system. They mate in the liver and many of the eggs get trapped in body tissues. In the immune response to this, surrounding tissue can be damaged. Before they infect humans, schistosomes infect snails. As it turns out, the Louisiana swamp red crayfish eats the snails. If the snails can be removed from the water, the schistosomes can't get to the point where they can infect humans. In experiments, ponds with established crayfish populations had greatly reduced snail populations, and schistosomes were less likely to be a problem. This type of biological control can be effective, reducing the need for chemicals, but it can be risky to introduce a new species.

Question to consider:

Explain an example of a biological control. See if you can think of (or find) an example other than the crayfish.

16. Life Application: Ecology and Human Conflict

How did ecological pressures contribute to the Rwandan genocide?

Jared Diamond, in his book *Collapse*, analyzes the ways that failed ecology contributed to the genocide in Rwanda.

Between April and July 1994 about 10% of the Rwandan population was murdered. This is generally attributed to ethnic hatred between the two major ethnic groups in Rwanda, the Hutu and the Tutsi, but that isn't the whole story because the genocide also occurred in Kamana region where only the Hulu lived. In 1991 the population density of Rwanda was 760 people per square mile, the highest in any African nation. It was 2040 people per square mile in the Kamana region.

Rwanda has good conditions for growing crops and is free of malaria and tsetse flies, so with the introduction of modern medicine, the habitat supported rapid population growth. Old farming techniques though, led to deforestation, overgrazing and nutrient depletion in soil plus loss of soil. With more and more people, farms got smaller and families got larger until the farms simply couldn't support the families. There was conflict between older established people with more land, and younger people. At least 5% of the Kamana population was killed. It wasn't all about ecology, but that was a contributing factor.

Question to consider:

How might an ecosystem failure lead to a conflict that could result in people getting killed?

16. How Do We Know?: Long-Term Ecological Research

What role do forests play in reducing water pollution?

It can take decades for change to occur in an ecosystem, so the National Science Foundation has established the Long-Term Ecological Research network of almost 2000 researchers at 21 sites throughout the United States. One of these is the Hubbard Brook ecosystem study in the White Mountains in New Hampshire. Since the 1960s, researchers at Hubbard Brook have been studying the nutrient and mineral cycling.

In one experiment, researchers compared two valleys for 3 years. They continuously removed all the plants from one, leaving them to decompose. The other was the control. As it turned out, in the control, plants took up 40% of the rain. All the calcium and potassium stayed in the forest, and nitrogen levels increased slightly. In the experimental valley, 100% of the rain left via the stream. There was a 10-fold increase in calcium loss, a 20-fold increase in potassium loss, and a 60-fold loss for nitrogen. So much nitrogen was added to the stream that algae increased to the point where the water was undrinkable.

Long-term studies like this confirm the role forests play in maintaining soil fertility and reducing water pollution caused by runoff.

Question to consider:

Why is it important to do long-term studies in ecology?

Student Study Guide

Chapter 17: Biodiversity and Human Affairs

How Is the Human Race Like a Meteorite?

Humans are having a large impact on the number of different species living on Earth: the biodiversity. This chapter looks at why this is, and why it matters.

17. Case Study: The Discovery of America

What role might humans have played in the extinction of many species thousands of years ago?

17. Case Study Summary

History credits Leif Ericson and Christopher Columbus for the discovery of America, but there were already people here when they arrived. For thousands of years, humans lived on Beringia, a low-lying area connecting what would become Asia and North America. The land was surrounded by glaciers but was kept warm by the Pacific ocean. Gradually, the climate was warming, and 18,000 years ago, these people ventured south in boats. They found wide coastal plains with ample food. They discovered America.

This population of a few hundred people colonized the Pacific coast of North America, and their population increased dramatically and expanded down the coast. By 15,000 years ago, humans had colonized the west coast of North and South America. By 12,000 years ago, they started moving inland. These people found elephants, camels, lions, horses, many kinds of deer, giant beavers, giant ground sloths, giant armadillos, rhinoceros, cheetahs, and dire wolves. All of these animals began going extinct about 11,000 years ago, and within 400–1000 years, they were gone (horses were reintroduced)— and at about that time, the human population expanded throughout North America. Coincidence?

Questions for Review:

Draw a timeline of human introduction and expansion in North America.

On your timeline, indicate when the extinctions of large animals began and when they ended.

List a few of the North American mammals that were around 12,000 years ago.

17.1 What Are the Components of Biodiversity?

Learning Objective 17.1

> *Distinguish between species diversity, genetic diversity, and habitat diversity, and describe how they relate to biodiversity.*

17.1   Summary

Biodiversity includes a number of parts:

- Species richness: The diversity of species alive in an area (how many different species there are). Higher species richness allows more varied interactions among species
- Genetic diversity: The degree of genetic difference among individuals in a species. In a genetically diverse species, the traits of individuals vary. When this is the case, there is a better chance that some of these traits will allow individuals to resist new diseases, tolerate changing climate, or face environmental challenges and the

species will not go extinct. Widely distributed species with large populations have higher genetic diversity. Thus, genetic diversity helps preserve biodiversity.
- Habitat diversity: Our planet contains many different types of habitats and different species are adapted to live in each, so more habitats leads to higher species richness overall. In addition, size matters. A bigger habitat supports more species than even a bunch of smaller ones.

All of these aspects must be maintained to conserve biodiversity. Often a species is endangered in the first place because its habitat or genetic diversity has been reduced.

17.1 Define these Key Terms

species richness

genetic diversity

habitat diversity

17.1 Questions for Review

17.1.1 What is meant by species richness? Can you think of a habitat or area on Earth that has high species richness?

17.1.2 Explain why it is better to be genetically diverse? Compare that to being more genetically uniform. What are the costs and benefits?

17.1.3 Draw a diagram illustrating a variety of different habitats on Earth. For each, draw or write in a few of the species that live there. Do they all have high species diversity? What happens to species diversity if you destroy all but two of them? What happens to species diversity only a small piece of each is left?

17.2 What Areas Have the Highest Biodiversity?

Learning Objective 17.2:

Explain the relationships between geographic relief, latitude, and biodiversity and the reasons why the latitudinal gradient exists.

17.2 Summary

Geographic relief refers to how much the terrain goes up and down. A mountainous area has high geographic relief. Because there are different habitats depending on altitude, there are different species live at each altitude. Habitats also differ depending on which direction they face. Thus high geographic relief tends to lead to higher biodiversity than in a flat plain.

The Latitudinal Gradient

As you move from the equator to the poles, the number of species decreases. The greatest diversity of species lives near the equator, and the least at the poles. In some cases, the tropics have more than 100 times as many species as temperate and polar regions. This is the latitudinal gradient and applies to almost every species.

Why the Gradient Exists

There are a few hypotheses, but we're not really sure.

- The climate hypothesis: The number of species declines as you move away from the equator because it gets colder, harsher, more variable, and less predictable. Of course some organisms do just fine, so why not others?

- The productivity hypothesis: The warm temperatures, sunlight, and rain in the tropics provide ideal conditions for plants, the base of the food web. In addition to the volume of food, the variety of plants allows animals to specialize and thereby share resources. Less variety at higher latitudes may require animals to be generalists, eating whatever they can, and this leads to greater competition and elimination of many species.
- The stability hypothesis: Compared with the coming and going of ice ages and glaciers at higher latitudes, the climate in the tropics has been more stable so species have simply had more time to evolve.

By far the most diverse places on Earth are coral reefs. They are mostly found in the tropics noted above, but they also have the features of mountain ranges that enhance diversity. They stretch from the cooler, darker, ocean floor to the warm and sunny surface, and they have a complex three-dimensional structure with many places to hide and vast surfaces on which to grow.

More than 80% of the world's species live in the tropical rainforests and reefs, and those are precisely the regions most threatened by human population growth, development, and climate change.

17.2 Define these Key Terms

geographic relief

latitudinal gradient

17.2 Questions for Review

17.2.1 Diagram a mountain. What will the habitat be like as you move higher? Wetter or drier? Colder or warmer? Trees? Grasses? Rocks?

17.2.2 The existence of the latitudinal gradient is well established. What is it?

17.2.3 Describe the three proposed hypotheses for explaining the existence of the latitudinal gradient.

17.2.4 Explain why a tropical coral reef is the most diverse habitat on Earth.

17.2.5 What are threats to tropical reef and rainforest habitats?

17.3 What Can Islands Tell Us About Biodiversity?

Learning Objective 17.3:

Describe the species-area relationship, and relate it to the theory of island biogeography.

17.3 Summary

Famous ecologists, R.H. MacArthur and E. O. Wilson, proposed the theory of island biogeography to explain why larger islands support greater species richness than smaller ones. There are two pertinent facts. First, every species on an island is an immigrant from somewhere else. Second, populations often go extinct on islands because they both start small and then have to compete for limited resources. Immigration increases the number of species and extinction reduces it. A bigger island closer to the mainland is going to have more immigrants and fewer extinctions than a smaller island farther away. Thus bigger islands have more species.

One result of this theory is the relationship between the area of an island and the number of species it has. As it turns out, this relationship takes the shape of a curve. A small island has many more species than a very small island, but a large island and a very large island have about the same number.

This theory of island biogeography applies to any ecological "island"—a small, isolated habitable habitat surrounded by unsuitable habitat. This could be a small patch of forest surrounded by houses. Just like any island, this small patch of forest can't hold all the species present in the original forest.

Many species on an island are endemic, or found nowhere else. More than 75% of the species of Hawaii are endemic—found only on those islands. Tropical islands are, therefore, a major source of diversity, and when those organisms go extinct, they're gone. They can't be brought in from somewhere else.

17.3 Define these Key Terms

theory of island biogeography

species-area relationship

endemic

17.3 Questions for Review

17.3.1 Draw a diagram of a mainland, larger inshore island, and smaller offshore island. Use it to explain the theory of island biogeography.

17.3.2 Based just on the discussion above, draw a typical species-area curve for a very small, small, large, and very large island (they should all be on the x-axis of a single graph)

17.3.3 From an animal's perspective, how is a small park in a housing subdivision like an island in the water?

17.3.4 Explain what an endemic species is. Why do you think human colonization of islands often leads to extinction of the endemics?

17.4 Why Do Different Regions Have Different Species?

Learning Objective 17.4:

Explain the relationship between Wallace's line and the distribution of species.

17.4 Summary

Planet Earth features multiple habitats: rainforests, deserts, grasslands, freshwater lakes, reefs, coastal intertidal zones, and more. There are organisms adapted to living in each of these, and probably something adapted to a desert in North America could survive in a desert in Africa, but you don't find the same species in those two deserts. Why?

Biogeographic Realms

Ecologists divide the planet into eight biogeographic realms, roughly corresponding to the continents. Each realm has its own set of species, although many species do occur in more than one realm.

A good example of how realms differ is found in New Guinea and Borneo in Southeast Asia. These islands are about the same size; they are both mountainous and covered with rainforest; and they have high biodiversity. Borneo is in the Indo-Malay realm and New Guinea is just 800 miles south in the Australia realm. And the difference in species is striking. Borneo has typical Asian animals: orangutans, rhinoceros, monkeys, woodpeckers. New Guinea has animals typical of Australia: cockatoos, tree kangaroos, many marsupials (the mammals that care for their young in pouches).

Wallace's Line

Alfred Russell Wallace was a naturalist in the 1850's, documenting the organisms in the Malaysian archipelago. He saw that his specimens fell into separate Asiatic and Australian groups, and the "line" separating them is now known as Wallace's line. (Be sure to review its location on Figure 17.11 in the text). The islands west of the line share the Asiatic fauna, and to the east they share the Australian fauna. The islands Bali and Lombok are only 20 miles apart, but on either side of the line, and they have completely different inhabitants. It turns out that geology explains this. The strait between them is deep with very fast currents, making migration between the islands difficult. And in the last ice age, Bali was part of the Asian mainland. When the glaciers melted and sea level rose, it became an island.

As the continents drift slowly around the planet, they can collide and connect, or be separate for long periods of time. Long separation allows for independent evolution of species which results in a unique collection for each biogeographic realm.

17.4 Define this Key Term

biogeographical realms

17.4 Questions for Review

17.4.1. Why is the fauna of the island of Bali so different than the fauna of Lombok, just 20 miles away?

17.4.2 Explain what Wallace's line is.

17.4.3 Why are species different from one biogeographic realm to the next? Or, why don't we find, say, cactus in deserts all over the world rather than just in North America?

17.5 How Does Biodiversity Change through Time?

Learning Objective 17.5:

Describe, in general, the variation in global biodiversity over the course of Earth's history.

17.5 Summary

Before 550 million years ago, all the species on Earth lived in the ocean. Most were unicellular. Multicellular animals were small, soft, and slow. Then the Cambrian Explosion happened and within 20 million years, every major group of animal we know now appeared in the fossil record. During the Cambrian period, from 542 to 490 million years ago, biodiversity exploded and ecologically complex communities of vertebrates, arthropods, mollusks and sea stars formed.

It's not clear why this happened when it did. Maybe increasing oxygen levels allowed larger and more active animals. Maybe hard body parts evolved at this time for protection from predators, and those fossilize in a way soft things don't. Whatever the cause, the Cambrian period is the beginning of global biodiversity.

17.5 Questions for Review

17.5.1 When did the Cambrian Explosion occur?

17.5.2 What is the Cambrian Explosion? What happened?

17.5.3 Why is the Cambrian Explosion significant in ecology and evolution?

17.5.3 Explain two hypotheses explaining the Cambrian Explosion.

17.6 Why Is Biodiversity Needed for a Healthy Ecosystem?

Learning Objective 17.6:

Detail how biodiversity enhances the productivity and stability of an ecosystem.

17.6 Summary

Conserving biodiversity costs money and can be controversial. Why worry about it?

Productivity, Stability, and Ecosystem Health

Productivity is an indicator of how effectively plants convert sunlight to food. It is measured by weighing the plant material produced in an ecosystem. This is where an ecosystem gets its energy, and ecosystem health is associated with higher productivity. (Depending on the type of ecosystem, of course. A perfectly healthy desert is gong to be less productive than a rainforest.)

The environment on Earth is variable and subject to disturbances: climate change, drought, fire. Ecosystem stability refers to the ecosystem's ability to resist these disturbances and return to its original state after them. A healthy ecosystem tends to be stable.

Why Biodiversity Increases Productivity

In every ecosystem studied (grasslands, forests, farms, coral reefs, rainforests, etc), higher productivity occurs with higher biodiversity. One reason is that a diverse collection of species can make more complete use of an ecosystem's resources. For example, a single species of plant is unlikely to be able to grow in all the conditions found in one ecosystem, but multiple plants can exploit them all. More plants means higher productivity. Also, some plants help other plants grow. Legumes add nitrates to the soil that other plants can use. Shrubs in deserts provide shade and moisture so seedlings of other species can get started. A more diverse ecosystem has more helpers, further increasing biodiversity.

Why Biodiversity Increases Stability

In a biologically diverse ecosystem, if a disruption affects one species, there is probably another that can step in and fill its role. If there are many species, at least some will do okay no matter what happens. A less diverse ecosystem doesn't have these redundancies and can't "bounce back" as quickly.

Biodiversity enhances ecosystem health in what's described as the rivet hypothesis by scientists Paul and Ruth Erhlich. The species in an ecosystem are like the rivets holding an airplane together. Some rivets can be lost, but lose too many, or particularly important ones, and the plane crashes.

How Biodiversity Keeps the Food Web Intact

Why don't herbivores eat all the plants and then die off? Is it plant defenses, or is it predators that keep the herbivore populations from getting too big? John Terborgh tested these hypotheses on islands in a large lake in Venezuela. The lake had been a valley, so each island has different species depending on what was there when the dam was completed and the valley flooded. On nine islands, there were no predators. On five, the normal mainland predators were present. As it turned out, without any predators, the islands were virtually stripped of vegetation. Predators maintained the ecosystem as normal on the other islands. So it turns out that biodiversity stabilizes the entire food web.

Sometimes a single species can have a disproportionate effect on biodiversity. These keystone species singlehandedly maintain the ecosystem. Sea otters off the Pacific Northwest coast of North America play this role in kelp forests. In the 1800's, sea otters were hunted for their fur, down to about 2000 animals. Sea otters eat sea urchins, which eat seaweed (kelp, in this location). Without sea otters to eat the urchins, the urchins decimated the kelp, eliminating the many other organisms that depend on kelp forests for food and shelter.

17.6 Define these Key Terms

productivity

stability

rivet hypothesis

keystone species

17.6 Questions for Review

17.6.1 How is ecosystem productivity measured, and explain how biodiversity increases that productivity.

17.6.2 Use the rivet hypothesis to explain how biodiversity increases ecosystem stability. What are some things that disrupt that stability and how does biodiversity allow an ecosystem to recover more quickly?

17.6.3 How do we know that the presence of predators is key to maintaining biodiversity? What happens if they're not present?

17.6.4 Using an example, explain what a keystone predator is.

17.7 Why Should We Preserve Biodiversity?

Learning Objective 17.7:

Evaluate the importance of old-growth forests for providing ecosystem services.

17.7 Summary

The Spotted Owl Controversy

In 1990, there were about 2000 northern spotted owls left in the United States and this species was put on the Endangered Species List. This owl lives only in the old-growth coniferous forests of the Pacific Northwest. Those big, 300-year-old trees were worth about $4000 per acre, and logging accounted for about half of Oregon's economy and about a quarter of Washington's. More than 10 million acres of old-growth forest was placed off limits to logging. Obviously, the families, towns, and industries who stood to lose that income weren't too interested in an owl. It wasn't really about the owl though. It was short-term versus long-term benefits.

Long-Term Benefits of an Old-Growth Forest

The spotted owl is an indicator species, meaning its presence indicates a healthy, functioning ecosystem. Old-growth forests in the Pacific Northwest are unique, temperate zone rainforests. They have very high biodiversity, like a tropical rainforest. Life exists from the crowns of the trees down hundreds of feet to the ground. It takes 300–1000 years for them to form.

Old growth forests provide valuable resources and services:

- The cancer drug *tamoxifen*, originally came from the Pacific yew tree from these forests. Odds are, other useful products are there too.
- They hold soil well in heavy rain, preventing erosion into lakes and rivers, which benefits salmon and other fisheries.
- Because water moves slowly through the forest, it gets filtered and cleaned, improving the quality of the area's drinking water
- They absorb a lot of carbon dioxide, helping to slow climate change
- Since they are very old, unique, and interesting, they support tourism

How Old Growth Forests Provide Ecosystem Services

You might remember the ecosystem services from Chapter 16:

- Provisioning services: Old-growth forests provides amazing genetic diversity with potential for drug development, and also some lumber.
- Regulating services: Old-growth forests retain and purify water.
- Cultural services: Old-growth forests support an eco-tourism industry and for many, provide a spiritual connection with nature.
- Supporting services: Old-growth forests are highly productive, create soil, and remove carbon dioxide from the atmosphere.

If the old-growth forests had been logged, they would have been gone in a decade or two, and logging would again be a financially strapped industry. In addition, critical ecosystem services would be gone. The people who profited would have done so at the expense of those who came after. And once biodiversity is lost and species go extinct, there is no way to bring them back.

The logging industry would have planted more trees but these manmade forests would be less diverse and therefore less productive, less stable, and would have required effort and money to maintain.

So why should we preserve biodiversity? To prevent long-term economic and ecological catastrophes. Because it is the ethical, or even spiritual thing to do. We now understand that we can affect biodiversity on a global scale, so preserving it may be a moral obligation as well.

17.7 Questions for Review

17.7.1 Explain why the logging industry in the Pacific Northwest was adamantly opposed to saving the northern spotted owl from extinction.

17.7.2 The federal government eventually decided to list the owl as endangered, requiring the protection of millions of acres of old-growth forest. Pretend you are one of the government workers charged with making this decision. Explain why you did it. List the pros and cons.

17.8 How Do We Keep Track of Biodiversity?

Learning Objective 17.8:

Describe how biodiversity is monitored.

17.8 Summary

Scientists have identified about 1.3 million species and are constantly finding more. Statistically, it's estimated that Earth could have 30 million species. Most of those we're missing are bacteria or other microscopic organisms, or insects and worms. Probably we know most of the mammals, birds, and plants. With so many species, it's difficult to characterize biodiversity, but it's important to be able to monitor and see changes. There are a few ways we do this.

The Species Diversity Index

Many ecological studies focus on a particular ecological community. There are three important features to measure to quantify biodiversity:

   o   Species richness: the diversity of species in an area
   o   Species abundance: the number of individuals of each species in the community
   o   Species evenness: the pattern of species abundance

Species diversity is highest when species richness is high and species abundance is fairly even. If there are a lot of just a few species, and just a handful of rare ones, like in a corn field, for example, the community would not be diverse.

A species diversity index is used to measure diversity. The mathematical formula takes in to account richness, abundance, and evenness. If a community's calculated diversity index is lower than expected, something is going on that is preventing species from living there: pollution, invasive species, drought. For example, the human migration across the Americas led to a decrease in the species diversity index.

Indicator Species and Satellite Images

It's important to monitor biodiversity globally, but species can't be counted on that large a scale. There are ways to approximate it, however:

   o   Indicator species: As long as the indicator species is okay, the ecosystem is okay. Birds and flowering plants work well for this because they are easy to survey.
   o   Satellite images: Satellites can collect photographs on a global scale that allows scientists to look for changes.

17.8 Define these Key Terms

species abundance

species evenness

17.8 Questions for Review

17.8.1 Describe the species abundance, richness, and evenness you would expect to see in a healthy, biodiverse ecosystem. What about one that is unhealthy?

17.8.2 Explain how the species diversity index is used to monitor biodiversity.

17.8.3 How is biodiversity estimated globally?

17.9 Why Might We Be Facing the Sixth Mass Extinction?

Learning Objective 17.9:

Compare the background and current rates of extinction, and explain how habitat destruction and overharvesting impact biodiversity.

17.9 Summary

The Cretaceous extinction, about 65 million years ago, was caused when a meteorite at least 20 kilometers across smashed into the ocean off the coast of Yucatan. It burned through the ocean, deep into the sea floor, sending huge clouds of steam into the air and 1000-foot tsunamis through the Atlantic. A shockwave of superheated air leveled and burned forests in North and South America. Millions of tons of ash blanketed the atmosphere, blocking light. This caused a complete collapse of global ecosystem services, and 75% of the species on Earth went extinct. This was the mildest of the "big five" mass extinctions in the history of the Earth. The largest extinction was the Permian, 245 million years ago, when more than 95% of species vanished.

The Blitzkrieg Hypothesis

In 1973, a scientist, Paul Martin, proposed that humans are a powerful force for extinction. This is the blitzkrieg hypothesis. One point in favor of it is that the large animals on every continent went extinct shortly after humans arrived, except for Africa, maybe because humans had been there longer. Donald Grayson, another scientist, argues that there is not much evidence that early humans hunted the early large animals that are now gone. Climate change might have killed them off.

Whoever is right, there is no doubt that human activity has driven many species extinct and threatens many more.

Background Extinction

Every species goes extinct eventually. The average lifespan of a species is between 1 and 5 million years. Around 99.9% of the species that have ever lived on Earth are extinct. Paleontologists use statistics to estimate extinction rates. Most of the time, there is a background extinction rate, much lower than the extinction rate during a mass extinction event. One common estimate of the background rate is one extinction per million species per year.

Current extinction rates are between 100 and 1000 times the typical background extinction rate. This is about on par with rates during past mass extinction events, thus we may be in a sixth mass extinction, driven by us. Right now, 13% of all bird species, 25% of mammals, 40% of amphibians, 28% of reptiles, 27% of fishes, and 71% of flowering plant species are in danger of extinction. If these rates are accurate, 60% of all species alive today will be extinct within 100 years.

Human Activity Threatens Biodiversity

The ecologist E. O. Wilson has quantified the ways human activity causes a decline in biodiversity. The percent of threatened or endangered species from each activity are as follows:

Habitat destruction	73%
Displaced by invasive species	68%
Chemical pollutants	38%

Overharvesting   15%

Many species are threatened by more than one of these so it adds up to more than 100%.

Habitat Destruction

This is most obvious and harmful in the destruction of the rainforests. These habitats account for 7% of the Earth's surface and contain more than 50% of the Earth's species. Satellite images allow us to track what happens. Once roads are built, logging starts. Villages grow up to support the industry. Roads expand. Farmers and ranchers move in and clear cut the forest for agriculture. Villages turn into small cities. Wherever this occurs, habitat destruction attacks all three components of biodiversity: species, genetic diversity, and habitat.

Overharvesting

Overharvesting occurs when individuals are taken from a population faster than they can grow back. It targets particular species that are valuable to us, but removing one species can destabilize and reduce the services provided by an entire ecosystem.

Marine ecosystems have been overharvested (whales, seals, sea otters, fish, crabs, lobsters). When this occurs, the first symptom may be a population crash and possible collapse of the harvesting industry. Conservation programs have brought many threatened species back, and we are learning how to harvest more sustainably.

17.9 Define these Key Terms

blitzkrieg hypothesis

background extinction

overharvesting

17.9 Questions for Review

17.9.1 The dinosaurs went extinct about 65 million years ago What happened?

17.9.2 What is the blitzkrieg hypothesis?

17.9.3 What is evidence for and against the blitzkrieg hypothesis?

17.9.4 Compare the background extinction rate with the current rate of extinction. What might this suggest about humans on Earth?

17.9.5 What are the four main ways humans affect biodiversity?

17.9.6 What leads to destruction of the rainforest, and why does that matter?

17.9.7 Explain what overharvesting is.

17.10 How Can We Preserve Biodiversity?

Learning Objective 17.10:

Examine the steps that can be taken to preserve biodiversity.

17.10 Summary

To preserve biodiversity we need to do the following:

- o Become educated about the long-term benefits: We need the services ecosystems provide

- o Develop economic policies that take advantage of long-term biodiversity benefits: We can "mine" ecosystems for new products, and set them aside for protection to reap the benefits of ecotourism.
- o Develop a cooperative approach: We need cooperation between scientists, politicians, economists, conservation organizations, organized religion, and informed citizens. International cooperation, especially to assist developing countries in avoiding overexploitation for short-term gain, is important.
- o Curtail human population growth: More people require more resources.
- o Design effective methods of conservation: Research on ecosystems and the organisms in them can help us design parks and preserves more effectively.

17.10 Question for Review

17.10.1 For each of the proposal above necessary to preserve biodiversity, write a sentence or two explaining how they would help us do this.

17. Biology in Perspective

Human activities are global in scale and we seem to be driving many other species extinct—like a meteorite. We're very good at competing for the resources we need, and other species lose out. Learning why biodiversity matters is the first key to changing this.

Species diversity, genetic diversity, and habitat diversity in stable ecosystems are critical to our survival. We depend on the services these ecosystems provide. In addition, life has intrinsic value, and we have a moral responsibility to be good stewards of the Earth for future generations.

Questions to consider:

Would you agree that humans are like a meteorite when it comes to biodiversity? Why or why not, and why does it matter?

17. Scientist Spotlight: E. O. Wilson

Why did E.O. Wilson make Time *magazine's list of the 25 most influential Americans?*

E. O. Wilson made Time magazine's 1995 list of the 25 most influential Americans. He published his first scientific paper at age 20. It was about invasive fire ants and he's studied ants ever since, publishing hundreds of papers and books as a professor at Harvard. He's won Pulitzer prizes and the National Medal of Science. This isn't really how he got on that list, though. He's become an ambassador for preserving biodiversity. He's seen firsthand the effects of habitat destruction globally and he understands and writes and speaks about the economic, cultural, and ethical benefits of preserving biodiversity.

Question to consider:

E.O. Wilson is about as famous as an ecologist can get. He's spending his later years focusing on getting the public interested in preserving biodiversity. What should this tell us?

17. Technology Connection: Satellite Imagery

How are satellite images being used to determine the rate of deforestation among the world's rainforests?

The U.S. Landsat program has been photographing global land cover by satellite since 1999. It records "photos" with seven different wavelengths of light, allowing composite images that are color coded to show different features on the land. It can distinguish roads, clearcuts, and larger buildings. As the satellite orbits and the Earth rotates, every place on Earth is photographed every 16 days. If we want to know something like how fast we're losing rainforest, we can clearly see and track deforested areas through time.

Question to consider:

How do we use satellites to monitor biodiversity?

17. Life Application: The Importance of Genetic Diversity

How could a jellyfish be used to stop Malaria?

Here is why genetic diversity is important to maintain. A certain jellyfish has a gene that makes a green fluorescent protein (GFP) that glows in the right light conditions. If this gene is inserted into a gene of interest in a study, that gene will glow when it is turned on and making proteins. Scientists insert this gene next to a gene active in larval testes in the mosquito that transmits malaria. This makes the male larvae glow green, and there is a machine that can isolate these males and sterilize them. This greatly reduces the number of malaria-carrying mosquitoes.

The scientists who discovered this glowing protein won the 2008 Nobel Prize for chemistry. Many of these glowing genes are now known and used in research. If that jellyfish had gone extinct before these guys came along, we might not have this useful tool. Even a single species can be economically and medically valuable.

Question to consider:

This is one example of a gene from some other organism that has turned out to be useful to us in some way. Can you think of other examples of such useful genes? Several have been mentioned in this text so far (remember genetic engineering?), but you may have heard of others as well.

17. How Do We Know: Experimental Island Zoogeography

Why did two scientists remove all of the species from an island off the Florida Keys?

Ecology requires field work for testing ideas. In 1969, Wilson and Daniel Simberloff designed an experimental test of MacArthur and Wilson's theory of island biogeography. They studied insects on six mangrove islands in the Florida Keys. They hypothesized that the number of insect species on each island would be relatively constant. One way to test this is to count the insect species on each island, remove them all and allow them to return. If the numbers of species were similar before and after, the theory would be supported. This is what they did. Imagine trying to count, and then remove all the insects on ten small islands! Overall, it turned out that the numbers before and after were similar, thus supporting the theory of island biogeography.

Question to consider:

Ecology has lots of mathematical models to explain processes in nature. Why is it important to test them in nature—the real world?

Student Study Guide

Chapter 18: Human Population Growth

How many people can a single planet hold?

18. Case Study: A Story About Bacteria

How much time do we have to address human population growth?

Like other species, humans are very good at reproducing. As populations grow though, habitat fills and overcrowding leads to starvation, disease, and predation. So far, agriculture, engineering, and technology have stretched our resources but continued growth is not sustainable. Luckily we have the ability to recognize the problem and work on ways to solve it.

18. Case Study Summary: A Story of Bacteria

Bacteria have the simplest growth possible. One cell divides in two, those two divide to produce four, then eight, and so on. Take a hypothetical bacteria species that divides once per minute. They live in a test tube culture and if one cell starts dividing at 11 am, the tube will be full of bacteria an noon. When does it become half full, the point at which the bacteria might be able to take steps prevent their extinction?

It turns out the tube is half full at 11:59. If it's doubling every minute, it will then be full at noon. So the bacteria decide to get 15 more tubes at 11:59. Since they're still doubling every minute, it takes 5 minutes to fill up all the tubes.

We are like these bacteria. Our habitat and resources are finite, and the earth can only absorb so much waste and pollution. We need to be sure we leave enough time to prepare so we need to understand how our population grows.

Questions for Review

Explain the parallels between bacteria growing in a test tube and the human population growing on earth.

What are the consequences of overpopulation of a habitat?

18.1 How Can Populations Grow So Fast?

Learning Objective 18.1

Distinguish between linear and exponential growth, and explain how the growth rate is determined.

18.1   Summary

Early on in human history, human population growth appeared to be linear. What does that mean?

The Difference Between Linear and Exponential Growth

In linear growth, a population increases at a constant rate. That's not how populations grow. Instead, as you see in the plot of human population growth (figure 18.2), the growth rate skyrocketed recently. This is exponential growth. This is because of compounding, where not only are new entities added to the population, the new entities can produce new entities. This is the case for any biological population, so they all have the potential for exponential growth.

Defining Growth Rate

Growth rate looks at how fast a population grows and how big it will get. If a village of 100 people grows by 3 in a year, that's a 3% growth rate. Population size multiplied by growth rate tells you how many individuals will be added in a particular unit of time.

Note that the number of people added to a population is proportional to the number already present. When the population is small, few people are added per year. As it gets larger, proportionally more people are added each year so growth accelerates and exponential growth is illustrated by the characteristic J-shaped curve. You can see how this curve can lull a population into a false sense of security. The population grows slowly for quite awhile, but then dramatically skyrockets.

Determining Growth Rate

When we say "growth rate," we really mean "net growth rate," taking into account that people are dying as well as being born:

> Growth rate = (birth rate – death rate)

Sometimes this is called the "natural" growth rate. In addition though, individuals migrate in and out of the population. The final form of growth rate, then, is:

Growth rate = (birth rate – death rate) + (immigration rate – emigration rate)

In human populations, these rates are expressed as numbers per 1000 individuals in a population (not per 100, which would give us percentages). These rates are shown for the U.S. in Table 18.1

Equilibrium

The growth rate is the 'balance" among birth, death, immigration, and emigration. If the positive components (birth and immigration) are greater than the negative, the population grows exponentially. If the negative components are larger, the population shrinks. If they are equal, growth rate is 0 and the population is in equilibrium with no change in size.

Doubling Time

In exponential growth, the amount of time required for a population to double is always the same. The doubling time for a population depends on the growth rate, not the population size. The constant 0.69 is used to determine doubling time:

> Doubling time = 0.69/growth rate

Doubling times are a bit easier to visualize and use than growth rates. In 2010 the growth rate for the U.S. was 8.3 per 1000 individuals. At that rate, the population would double in 83 years. This information is useful for future planning for facilities or resource requirements.

18.1 Define these Key Terms

linear growth

exponential growth

compounding

growth rate

equilibrium

doubling time

18.1 Questions for Review

18.1.1 Sketch an approximation of the growth curve for the human population (don't worry about numbers). Explain why it looks this way?

18.1.2 On your growth curve, do you see where one could conclude that the growth rate was linear early on? How would a linear growth curve be different in outcome than the exponential curve (you don't see linear growth curves in populations, but you do see them in other things)

18.1.3 There is a population of 10,000 people and last year, 40 people were added. What's the growth rate?

18.1.4 There is another population of 5000 people. Last year, 10 were born, 3 died, 3 moved in from elsewhere, and 1 moved elsewhere. What's the growth rate?

18.1.5 What if, in our population of 5000, 10 people were born, 3 died, none moved in, and 7 moved out? What's the growth rate?

18.1.6 Calculate the doubling time for both the population of 10,000 and the population of 5000 (from 18.1.5 above).

18.2 Why Don't Populations Grow Forever?

Learning Objective 18.2:

Describe the effects of population density and carrying capacity on the principle of logistic growth.

18.2 Summary

Clearly we would notice if populations living on earth grew exponentially without stopping. What stops it?

The Effects of Population Density

An increasing population eventually has negative impacts on its environment and members. Food and space becomes inadequate. Waste products increase. Diseases and parasites spread more easily. Crowding leads to conflicts. These are density dependent factors: they get worse with increasing population density. They make it more difficult to reproduce and survive. Is this happening to humans?

In the 18th century, economist Thomas Malthus first suggested that humans were in trouble. He pointed out that while our population was growing exponentially, our food supply grew linearly. But global production of food has so far kept up with population growth due to the development of high yield crops and productive, technology dependent agriculture.

Logistic Growth

Density dependent factors are not taken into account in exponential growth, which works for awhile, with birth rates far exceeding death rates. At some point though, the population gets too big and life gets hard. Birth rates decline and death rates increase. As a result, growth rates slow. At some point birth and death rates become equal and the population is at equilibrium or its carrying capacity. Carrying capacity is the maximum number of individuals the environment can support.

With density dependent factors taken into account, we get a new pattern of population growth: logistic growth. Here growth starts off fast but declines and eventually stabilizes at the carrying capacity. It's an s-shaped curve. This is generally what happens in nature. If a population exceeds carrying capacity, individuals will starve and die until it returns to equilibrium.

Globally, the human population doubled between 1960 and 2000. If we continue to double every 40 years, we are likely to run out of food and drinking water. Already 11% of the world population lacks dependable drinking water. Removing wastes and pollution and providing education and healthcare would become impossible.

Luckily, it appears that global population growth rates are finally slowing down, trending toward the logistic phase. Current growth rate is about 1.3% annually and projected to be about 0.5% by 2050. At that rate we will reach 9 billion by 2015 and level off near 10 billion by 2200.

Our Carrying Capacity

Since carrying capacity for humans depends on technological breakthroughs and quality of life, it's difficult to predict. Estimates run as low as 2-3 billion, but generally fall between 8 and 15 billion. It's good that our growth rate does seem to be slowing, but a 50% population increase by 2050 is going to be difficult to deal with, and many developing countries, which already can't sustain their populations, continue to have high growth rates.

18.2 Define these Key Terms

density dependent

carrying capacity

18.2 Questions for Review

18.2.1 List some density dependent factors which would tend to slow population growth. How is it they slow growth?

18.2.2 In the 18th century it seemed pretty clear we were going to outstrip our resources on earth. We're still here though, so what happened?

18.2.3 Sketch a logistic growth curve (don't worry about numbers). Where is the carrying capacity?

18.2.4 What's the prognosis for human population growth in the future? Are we doomed?

18.3 How Is Population Growth Influenced by Age and Sex?

Learning Objective 18.3: How Is Population Growth Influenced by Age and Sex?

Explain what age pyramids can tell us about a country's conditions and social challenges.

18.3 Summary

All of the growth models we've looked at so far assume that all individuals reproduce equally. This, of course, is not the case.

Age, Sex, and Population Growth

A population can only grow as fast as females produce children, so adding or subtracting females more directly affects population growth than males. Females can only produce children for part of their lives though, and they vary in their fertility.

Age affects growth rate in general, because it alters the chance of survival. A disease of older people may not affect reproductive rate or population growth. A disease of children will affect future growth. A disease that affects women in their reproductive years can have a huge impact on population growth.

Age Pyramids

The science of demography analyzes population growth based on its age structure. An age pyramid illustrates this data for a population at a given point in time. It is essentially a sideways bar graph with males and females on either side. The bars on either side show the number of individuals in a given 5-year age class. There are three typical shapes:

- o Widest at the bottom: This is typical because normally children are the largest age group and the pyramid narrows as people age and die.
- o Reasonably vertical sides: In this case there are approximately equal numbers of each age group
- o A top wider than the bottom: There are more older people than younger

As an example of how this is a useful too, if you look at the age pyramids for the 1950's through the present, you can trace a large bulge, moving from bottom to top, which represents the baby boomers. This has had important consequences. Early on we needed schools and teachers. In their middle years, the boomers paid a lot of taxes. As they retire, spending on health care and social security has increased. The age pyramids also show that boomer mortality is fairly low, and that they have produced a mini "boomlet" of children. They live long, and women live longer than men. By looking at the age pyramids, one could predict these social needs and patterns.

Age pyramids are useful for comparing growth patterns of different countries.

18.3 Define these Key Terms

demography

age pyramid

18.3 Questions for Review

18.3.1 Who has more of an impact on population growth rates, and why? Males or females?

18.3.2 Sketch the three types of age pyramids. Don't worry about numbers of people, but do include ages of the people on your graphs. For each, describe what is going on with that population.

18.3.3. For one of your graphs, add a large bulge of children and then sketch a new pyramid for this population every 20 years. Do you see how you are able to predict subsequent pyramids once you see the first one in the series?

18.3.4 For each of your series of pyramids in 18.3.3, explain how the bulge may be affecting society that year.

18.4 Why Do Developing and Developed Countries Grow Differently?

Learning Objective 18.4:

Compare and contrast the factors that contribute to growth in developing and developed countries.

18.4 Summary

There are two additional demographic factors that affect population growth: total fertility and the age at first reproduction.

Total Fertility and Age of First Reproduction

Total fertility is the average number of children a woman could have. Birth rate is the average number of children a woman does have. Since not all women have all the children they can, total fertility is higher than birth rate and gives a more accurate picture of a population's potential to grow. If total fertility is 2, each couple is replacing itself and there is zero population growth (ZPG). More than 2 and the population increases. Less than 2 and it decreases.

Age at reproduction (not age of puberty) is important because the earlier women have children, the more they can have. Women in developed countries tend to have children later, reducing population growth in two ways. One, fewer children, and two, fewer generations alive at the same time. If people routinely know their great-grandparents, that's 4 generations alive at once.

Fertility and Mortality Differences

Total fertility of developing countries is much higher than developed countries. Couples chose family size based on economic, religious, and social considerations, but in developing countries it comes down, in part, to infant mortality. In developing countries infant mortality is very high so having a lot of children ensures that at least some will survive. This has to do with malnutrition, disease, lack of sanitation, and limited medical care. These are density dependent factors, and, indeed, population density of developing countries is much higher than in developed countries.

Family Planning Differences

Biologically, many girls can become pregnant before age 13. In some countries, puberty is considered the age of consent for marriage and children follow quickly thereafter. If life expectancy for parents is short, this provides security for a daughter, and also gives the parents one less mouth to feed. Sometimes girls marry early for religious reasons. Sometimes these women are denied a right to an education, and a right to vote, and are kept subservient to their husbands, so they lack access to contraception and have limited financial and social independence.

Early marriage increases total fertility and lowers the age of first reproduction, both of which increase population growth. This is somewhat mitigated because many of these young girls die in childbirth.

In developed countries, women have greater access to family planning and are more likely to be in a position to postpone childbearing because of education or a career. Total fertility is reduced. Age at first reproduction is increased. Population growth decreases.

It turns out that women's rights are perhaps the most significant difference between developing and developed countries, and consequently on their population growth rates. Education has a particularly powerful influence on reproduction.

Demographic Transition

Over time, industrialization occurs, urbanization increases, and education becomes more important as children become a less important part of the labor force. Developing countries become developed. This results in a demographic transition:

- o Pre-development: high birth and death rates, the population lives close to carrying capacity, and overall population growth is limited
- o Transition: improved agriculture and interactions with other countries, resulting in better medical practices, more sanitation, cleaner water supply, more technology, and better education, which increases carrying capacity. Death rates fall and life expectancy increases. Population growth increases exponentially.
- o Post-Transition: birth rates drop to match death rates, urbanization increases, the status of women improves, there is access to contraception and women postpone childbirth. Population growth levels off and eventually stops. The population may actually shrink.

Most of our presently developing countries are in the transition stage so their populations are growing exponentially. Investing in their economies, medicine and technology in order to slow their population growth is worth doing to slow this down.

18.4 Define these Key Terms

total fertility

demographic transition

18.4 Questions for Review

18.4.1 Explain how total fertility and age of first reproduction work together to increase population growth rates.

18.4.2 Women in developed countries have children later. How does this affect population growth rates? (there are two factors at work)

18.4.3 Why does higher infant mortality end up increasing population growth? And why do developing countries tend to have higher infant mortality?

18.4.4 Giving women a right to an education is the single most effective thing to do if you want to reduce population growth rates. Explain the reasons for this, listing as many as you can.

18.4.5 Diagram a time-line of the demographic transition of a hypothetical country. List the events and the relative population growth rate at each stage.

18.5 How Do We Use Information About Population Growth?

Learning Objective 18.5:

Demonstrate how data on population growth can be used to plan for population shifts and resource depletion.

18.5 Summary

Demographic information is useful to many groups:

- o Insurance companies predicting lifespans
- o Market researchers deciding what to sell to whom
- o The CIA because population pressure poses a potential security threat
- o The government for organizing representation

Anticipating population growth and demographic shifts allows us to plan for society and analyze our use of resources.

The Constitution and the Census

A representative government requires detailed knowledge of the people being represented, so the U.S. Constitution mandates a census at least every 10 years. Today the census counts individuals and keeps records of age, sex, ethnicity, family size, marriage, divorce, employment status, household income, and more. These data help Congress allocate funds for public utilities, roads, and schools.

Planning for Population Shifts

Demographers can look at the present age pyramid and predict what it will be in the future, or look at the history. Uganda's age pyramid (Figure 18.11) reveals low rates of survival for all age classes. The country has endured the purges of a dictator and a civil war. Up to half a million people were killed directly and many more died from starvation, lack of medical care or other social services, and the AIDS pandemic. The wide base of the pyramid now indicates a high birth rate and the early stages of a demographic transition.

The projected age pyramid for Uganda in 2050 (Figure 18.17) shows improved survivorship and a much larger overall population. By 2050 the population size will have doubled and half those will still be children. Uganda is going to have to focus on the needs of childhood: nutrition, education, controlling diseases, and eventually, providing jobs.

Italy, on the other hand, has an aging population projected to decline by 2050. It will need to focus on caring for the elderly: cancer, heart disease, and palliative end of life care, in addition to figuring out how to increase worker productivity as there will be fewer workers to go around.

Resource Depletion

Finite resources cannot support an exponentially growing population indefinitely. We use oil for transportation and heat, and in a huge variety of products. For the last 30 years, the U.S has used about ¼ of the world's annual oil production. As developing countries develop, they will demand more oil. Scientists debate how much oil is left to be extracted, but we can't assume that because it's taken us 150 years to use half our oil, the rest will last another 150 years. We'll consume that oil in one doubling period, or about a decade or two.

We can reduce oil consumption. The cost of getting more and more remote reserves may drive consumption down. Technology can improve efficiency and identify replacements for oil. Oil consumption is down in the U.S.

The Limits in Growth

The planet cannot support an ever-increasing human population. But we know this and so there is hope that we will be able to transition from growth to global equilibrium before it is too late.

18.5 Questions for Review

18.5.1 List a few examples of how demographic information can be useful to various groups.

18.5.2 Why does the U.S Constitution mandate a census? What good is that information to a government?

18.5.3 Sketch the present age pyramids for Uganda and Italy. Compare the consequences of these for the two societies. What do they predict and so what do these countries need to plan for?

18.5.4 How does population growth rate relate to resource depletion rate? What eventually happens? How can this problem be solved?

18. Biology in Perspective

The human population on earth will reach 10 billion people by 2200. We may hope that we're at the transition point where we stop growing exponentially, growth slows and becomes logistic.   Demographics let us predict population trends to guide planning and resource allocation, and demographic changes slow population growth in developing countries. Even developed countries, though, use up essential resources. Density dependent factors may end up reducing our growth rate by disease, starvation, and insufficient resources, but we are in a position to make choices to limit our growth ourselves.

Question to consider:

Explain how population growth eventually limits itself. Why might that be a bad thing for humans?

18. Scientist Spotlight: Donella Meadows

In what way did Donella Meadows' work pose a threat to conventional thinking about the environment?

Dr. Donella Meadows studied environmental problems analytically and qualitatively using "system dynamics," where a computer can simulate change in a complex system. She and Jay Forrester were able to study how factors like resource use, medical advances, and pollution affect population growth, and how population growth affects them. They published their findings in a book called The Limits to Growth. It had a disturbing conclusion

If the present growth trends in world population, industrialization, pollution, food production, and resource depletion continue unchanged, the limits to growth on this planet will be reached sometime within the next hundred years. The most probable result will be a rather sudden and uncontrollable decline in both population and industrial capacity.

This argued for sustainable living rather than the conventional model that depends on steady economic growth.

Question to consider:

In the U.S., steady economic growth seems to be necessary. If this is the model, why would what Donella Meadows proposed be so controversial?

18. Technology Connection: Male Contraception

How did scientists studying deafness discover a clue that could lead to effective male contraception?

For contraceptives, males have three options: condoms, vasectomies, and abstinence. Generally they've left this to the women to deal with, but there are some promising lines of research.

1. Testosterone plays an important role in sperm production, but too much testosterone actually reduces it. The hormone could be used somehow to reduce male fertility, although it's not entirely clear how this works and testosterone does other things too, so altering levels is risky.

2. Scientists were doing a genetic analysis of Iranian families that had a high incidence of deafness. They also found that in two families, male infertility seemed to be inherited. DNA analysis showed a mutation in a gene called CATSPER1, which makes the protein found in the tail of sperm cells. The tail has to wiggle violently to get the sperm through an egg's membrane for fertilization. The CATSPER1 mutation prevents this. Maybe if we could knock out CATSPER1 in males, they would not be able to father children. This could be done by immunocontraception where antibodies that attack the protein are simply injected. Fertility returns when the injections stop.

Question to consider:

What, in general, is immunocontraception?

18. Life Application: The Demographics of China

Why are China's population control measures so controversial?

China has a fifth of the world's population, so they have been especially concerned about problems associated with high population growth and instituted a one child per couple policy in the 1970's. By some measures, this has been successful. The Chinese population is expected to peak at 1.46 billion around 2030 and then decline. Despite this success, government-imposed restrictions on reproduction have been very unpopular. It has definitely led to a preference for boys over girls. Among adults now, there are 27 million more men than women.

Question to consider:

China's one child per couple policy has held down their population growth rate rather successfully. Do you think this would be a reasonable approach for the U.S.?

18. How Do We Know: Modeling Population Growth

How can we determine how big the world's population will be in 50 years?

We can predict the size of our future population with a few calculations.

First, how fast is the population growing? The number added to the population is growth rate, r, times population size, N.

$$\text{Number added} = rN$$

We add the number of people added in a year to N to see how much the population grew in a year. To get to the population in 50 years, you also determine the number of people added in year 2 and year 3 and so on, and then add all of them together. There is an equation that does this:

$$\text{Size} = N_0 e^{rt}$$

N_0 is the starting population size, r is growth rate, and t is the number of years in the future. Raising e to a power is called the exponential function, which is where exponential growth gets its name. This lets you predict the population into the future, and then other calculations take in to account the density dependent factors that will affect population growth.

Question to consider:

Why is it important to be able to calculate population into the future?